Electrostatic Phenomena on Planetary Surfaces (Second Edition)

Online at: https://doi.org/10.1088/2514-3433/ac878e

AMERICAN ASTRONOMICAL SOCIETY | IOP ebooks™

About the program:

AAS-IOP Astronomy ebooks is the official book program of the American Astronomical Society (AAS) and aims to share in depth the most fascinating areas of astronomy, astrophysics, solar physics, and planetary science. The program includes publications in the following topics:

- GALAXIES AND COSMOLOGY
- INTERSTELLAR MATTER AND THE LOCAL UNIVERSE
- STARS AND STELLAR PHYSICS
- EDUCATION, OUTREACH, AND HERITAGE
- HIGH-ENERGY PHENOMENA AND FUNDAMENTAL PHYSICS
- THE SUN AND THE HELIOSPHERE
- THE SOLAR SYSTEM, EXOPLANETS, AND ASTROBIOLOGY
- LABORATORY ASTROPHYSICS, INSTRUMENTATION, SOFTWARE, AND DATA

Books in the program range in level from short introductory texts on fast-moving areas, graduate and upper-level undergraduate textbooks, research monographs, and practical handbooks.

For a complete list of published and forthcoming titles, please visit iopscience.org/books/aas.

About the American Astronomical Society

The American Astronomical Society (aas.org), established 1899, is the major organization of professional astronomers in North America. The membership (~7,000) also includes physicists, mathematicians, geologists, engineers, and others whose research interests lie within the broad spectrum of subjects now comprising the contemporary astronomical sciences. The mission of the Society is to enhance and share humanity's scientific understanding of the universe.

Editorial Advisory Board

Electrostatic Phenomena on Planetary Surfaces (Second Edition)

Carlos I Calle and Karen L Aplin

IOP Publishing, Bristol, UK

ISBN 978-0-7503-3891-2 (ebook)
ISBN 978-0-7503-3889-9 (print)
ISBN 978-0-7503-3892-9 (myPrint)
ISBN 978-0-7503-3890-5 (mobi)

DOI 10.1088/2514-3433/ac878e

Version: 20221001

AAS–IOP Astronomy
ISSN 2514-3433 (online)
ISSN 2515-141X (print)

British Library Cataloguing-in-Publication Data: A catalogue record for this book is available from the British Library.

Published by IOP Publishing, wholly owned by The Institute of Physics, London

IOP Publishing, No.2 The Distillery, Glassfields, Avon Street, Bristol, BS2 0GR, UK

US Office: IOP Publishing, Inc., 190 North Independence Mall West, Suite 601, Philadelphia, PA 19106, USA

Cover image: Shallow Lightning on Jupiter (Illustration).
Image credit: NASA/JPL-Caltech/SwRI/MSSS/Gerald Eichstädt.

To my grandson, Liam Daniel Calle

Carlos I Calle

To the memory of my aunt and uncle Linda (1950–2022) and Norman Neal (1949–2021), both scientists taken too soon by Covid

Karen L Aplin

Contents

Preface to the First Edition

Our knowledge of planetary environments has increased considerably as a result of the planetary exploration missions that NASA primarily along with ESA and other space agencies have launched in the last decades. Electrostatic phenomena occurring on the solar system bodies are among the most important to consider, as they affect the weather in planets and moons with atmospheres and the physical properties of the surfaces on planets and moons lacking an atmosphere. These phenomena are not completely understood even on Earth. Data on electrostatic phenomena from recent missions are currently being analyzed and results are being published and discussed in scientific journals and at conferences.

My aim with this concise book is to provide an overall understanding of the different aspects of electrostatic phenomena as they occur on planetary atmospheres and surfaces, to show the reader what is known in this important field, what is expected from planned exploration missions, and what the big unknowns are. I have included numerous references at the end of each chapter to guide further research into this field.

Preface to the Second Edition

For this new edition, we have added a new chapter on space radiation that includes the types of radiation, how they are measured, their effects on materials and humans, and how radiation can be mitigated. We have also moved the sections on the electrostatic environment of Jupiter into a separate chapter. We have updated the book throughout, with data from recent missions. Space science and technology research moves rapidly, and this second edition includes new results and analysis from missions and experiments.

Acknowledgements

The idea for this concise book on electrostatic phenomena on planetary surfaces came from Nicki Dennis at Morgan & Claypool and Institute of Physics (IOP) joint publishing. Dr. Leigh Jenkins at IOP Publishing provided guidance for the second edition. We are indebted to them. We would also like to thank Dr. José Nuñez at NASA Kennedy Space Center for his support of this work.

Dr. Calle would like to acknowledge the valuable contributions to the work on electrostatic phenomena on the Moon and Mars by the research team at the NASA Kennedy Space Center Electrostatics and Surface Physics Laboratory. Their work has led to the development of several electrostatics-based technologies for NASA's planetary exploration missions. He would like to acknowledge in particular Dr. Charles Buhler, Dr. Sid Clements, Paul Mackey, Dr. Michael Hogue, Dr. James Mantovani, Michael Johansen, James Phillips III, Jerry Wang, Dr. Michael Dupuis, Dr. Aaron Olson, Dr. Joseph Toth, Dr. Krystal Acosta, and Seth Rhodes. He would also like to thank Dr. Luz Marina Calle, his wife and fellow NASA scientist. She has always been a wonderful sounding board during his research and writing endeavors.

Prof. Aplin would like to thank collaborators past and present who have provided many stimulating conversations and ideas relevant to this book, particularly, Prof. Giles Harrison, Prof. Galina Bazilevskaya, Prof. Lucy Berthoud, Prof. Neil Bowles, Dr. Georg Fischer, Barry Kent, Prof. Mike Lockwood, Prof. Michael Rycroft, Prof. Chris Scott, Prof. Yukihiro Takahashi, and Prof. Yoav Yair. Technical support is a vital part of developing new instrumentation and help with design and manufacturing from Adam Baird, John Firth, Peter Hastings, Jeff Lidgard, and Keith Long is very much appreciated. Finally, the enthusiasm and creativity of numerous under- and post-graduate students is also acknowledged.

Author Biographies

Carlos I Calle

Carlos I Calle is founder of the Electrostatics and Surface Physics Laboratory at NASA Kennedy Space Center. He works on the problem of electrostatic phenomena of granular and bulk material as they apply to planetary surfaces, particularly that of Mars and the moon, developing instrumentation for future planetary exploration missions. He has over 180 scientific publications, five books, and several patents.

Dr. Calle received the NASA Exceptional Technological Achievement Medal in 2010, the NASA Spaceflight Awareness Award in 2003 for his outstanding contributions to the space program, the NASA Silver Snoopy Award in 2007 for his exceptional contributions to human spaceflight and the Kennedy Space Center Director's Innovation Award in 2022 for his support of NASA's Artemis mission.

Other books by Carlos I. Calle:

Superstrings and Other Things: A Guide to Physics (3rd ed.; London: CRC Press/Taylor & Francis), 454 pp. (2020)
  French edition: *Supercordes et Autres Ficelles: A Voyage au Coeur de la Physique* (Paris: Dunod), 608 pp. (2004)
The Universe, Order without Design (Amherst, NY: Prometheus Books) (2009)
Coffee with Einstein (London: Duncan Baird Publishers), 150 pp. (2007)
  Greek edition: *Συζητώντας με τον Αϊνστάιν* (Agia Paraskevi: Dioptra) (2021)
  Slovakian edition: *Rozhovory s Albertom Einsteinom* (Bratislava: Ikar Publishing House) (2021)
  Japanese edition: アインシュタインとコーヒー (Tokyo: Sangnesha Publishers, Inc) (2015)
  Turkish edition: *Einstein, Hayali Söyleşiler* (Itstanbul: Kolektif Kitap) (2012)
  Chinese edition: 在咖啡馆遇见爱因斯坦 (Beijing: Wang Zheng Heilongjiang University Press) (2013)
  German edition: *Auf einen Kaffee mit Einstein* (Munich: Deutscher Taschenbuch Verlag) (2009)
  French edition: *Un café avec Einstein* (Paris: Gründ) (2008)
Einstein for Dummies (Hoboken, NJ: Wiley), 374 pp. (2005)
  Dutch edition: *Einstein voor Dummies* (Amsterdam: Addison-Wesley/Pearson Benelux B.V.), 364 pp. (2005)
  Spanish edition: *Einstein para Dummies* (Bogota: Grupo Editorial Norma) (2006)
  Serbian edition: *Ainštajn za neupućene* (Banja Luka: Mikro knjiga) (2006)

Karen Aplin

Karen Aplin is Professor of Space Science and Technology in the Faculty of Engineering, University of Bristol, UK. Her research work includes ionization, space weather, novel instrumentation, and recovering historical electrostatic data. She is a founding editor of the academic journal *History of Geo- and Space Sciences*. Previous appointments include at the Rutherford Appleton Laboratory and the University of Oxford. Professor Aplin has published around 100 peer-reviewed papers and two books. She is a regular contributor to print and radio media, and a BBC Radio 3 documentary was centered on her work on weather in classical music.

Professor Aplin received the James Dungey Lectureship of the Royal Astronomical Society in 2021 for her work on planetary atmospheric electricity and a European Space Agency Certificate of Outstanding Achievement in 2017 for contributions to the Rosetta mission. She is a Fellow of the Institute of Physics and the Royal Astronomical Society.

Other books by Karen Aplin:

Electrifying Atmospheres: Charging, Ionisation and Lightning in the Solar System and Beyond (Netherlands: Springer), 57 pp., ISBN 9789400766334, doi:10.1007/978-94-007-6633-4 (2013)

F. Leblanc, K. L. Aplin, Y. Yair, R. G. Harrison, J. P. Lebreton, and M. Blanc (ed.) *Planetary Atmospheric Electricity* (New York: Springer), ISBN 978-0-387-87663-4, doi:10.1007/978-0-387-87664-1 (2008)

Carlos I Calle and Karen L Aplin

Chapter 1

Introduction

The diverse planetary environments in the solar system react in somewhat different ways to the encompassing influence of the Sun. These different interactions define the electrostatic phenomena that take place on and near planetary surfaces. The desire to understand the electrostatic environments of planetary surfaces goes beyond scientific inquiry. These environments have enormous implications for both human and robotic exploration of the solar system.

In solar system bodies with an atmosphere, electrostatic phenomena are determined by the physical and chemical properties of those atmospheres. Atmospheric pressure and density, as well as composition, govern the way objects acquire, hold on to, and release electrostatic charge. The transfer of charge from an object to another can happen in different ways. If the transfer is fast, as in electrical breakdown, lightning can occur. In addition to Earth, lightning has been detected on Jupiter and Saturn and there are disputed observations of it at Venus. Atmospheric models predict that lightning should also occur on Mars, and on Saturn's moon Titan. Despite several efforts to detect lightning on these bodies with instrumentation aboard spacecraft, the question remains open. Electrostatic discharges have been detected from Uranus and Neptune, although it is not clear if those discharges are caused by lightning.

When the charge transfer is slow, a glow discharge known as a corona can take place. These corona discharges occur on Earth and may also occur on Venus, Mars, Titan, and the giant planets.

The most extensively studied electrostatic environment is clearly that of the Earth. The Earth's magnetosphere, produced by the interaction of the solar wind with the planet's magnetic field, controls the electrostatic charge content of the atmosphere. The magnetosphere slows down, deflects, and traps many solar wind particles. These trapped particles form the doughnut-shaped regions called Van Allen radiation belts, a nearly impenetrable barrier that prevents the most energetic electrons from reaching the Earth's atmosphere.

doi:10.1088/2514-3433/ac878ech1

Spacecraft and satellites interact with the space plasma environment around the Earth. This interaction generates electrostatic charging on these orbiting craft, a complex phenomenon that may interfere with their operation and may disrupt or damage power, navigation, communications, and other instrumentation.

In addition to the Earth, Mercury, Saturn, Jupiter, Neptune, and Uranus, as well as Jupiter's moon Ganymede have magnetic fields. As happens on Earth, these magnetic fields interact with the solar wind, producing magnetospheres that act as semipermeable barriers to the solar wind particles. The magnetic field lines plunge at the planets' magnetic poles, allowing particles from the solar wind to reach lower altitudes. Mass ejections from the Sun's corona, commonly known as solar flares, increase the flux of solar wind particles for several hours. These particles interact with the magnetic fields of the planets, releasing trapped particles, which trigger reactions with atmospheric molecules that release photons. In the polar regions, these photons form the auroras, known as the northern and southern lights on Earth. Auroras have been observed on Jupiter, Saturn, Uranus, and Neptune.

The electrostatic environment of airless bodies, such as the Moon, Mercury, and the major asteroids, is the result of the direct interaction of the surface with the solar wind, cosmic rays, and solar radiation. The surfaces of these solar system bodies develop a charge that balances the sum of all these current fluxes. The electrostatic interaction between the charged surfaces of these solar system bodies and the surrounding plasma results in the arrangement of the plasma particles in the form of a shield that surrounds the surface. This shield limits the magnitude of the charge that develops on these surfaces.

The surfaces of Mars, Mercury, the Moon, and a few other solar system bodies, are covered with a layer of fine dust. Interaction of the surface dust with the robotic exploration rovers on Mars or with rovers and astronaut boots during missions to the Moon generates electrostatic charge on the bodies as they repeatedly make contact and separate, a phenomenon called triboelectric charging. Triboelectric charging is unlikely to be a concern on the day side of airless bodies, but can reach levels that may cause electrostatic discharges on the dark sides due to a stronger electron flux in those regions. Triboelectric charging is also a concern on Mars, and NASA took steps to mitigate the issue with all its rovers operating on the planet.

Data from recent and current planetary missions from NASA, the European Space Agency, and the Japanese space agency JAXA has provided unprecedented information on the electrostatic environments of solar system planets and moons. Some of these data are still being analyzed and new discoveries are frequently made. But many unknowns remain. Not all planned experiments in planetary missions have been successful. Some missions carrying valuable experiments have failed altogether, particularly to Mars, which is a difficult destination to land on.

In the pages that follow we describe what is known about the electrostatic environment of the solar system from early and current experiments on Earth, as well as what is being learned from the instrumentation on the space exploration missions of the last few decades. But before embarking in this study, we present a brief review of the basic principles of electrostatics.

Chapter 2

Electrostatics Principles

A brief introduction to the fundamental principles of electrostatics may be of use before looking in more detail at the electrostatic phenomena on the different bodies of the solar system.

Electrostatics is a component of electromagnetism, a branch of physics that describes the behavior of the electromagnetic force, one of the four fundamental forces in nature. The fundamental problem of electromagnetism is to understand the force that electric charges exert on one another. Electrostatics restricts itself to answer this question when charges are not in motion. As we shall see, electrostatics does allow for short, slow motions of charges as they interact with each other. Continuous motion of charges generates magnetic fields and that takes us beyond electrostatics to the range of electromagnetism.

2.1 Coulomb's Law and the Principle of Superposition

The solution to the fundamental problem of electromagnetism was given by Charles Coulomb in the eighteenth century and is known today as Coulomb's law. If two charges q_1 and q_2, considered to be *point* charges (that is, charges on an ideal object with no physical extension), are separated by a distance r, the force that q_2 exerts on q_1 is

$$\mathbf{F} = k\frac{q_1 q_2}{r^2}\hat{\mathbf{r}} \tag{2.1}$$

where $\hat{\mathbf{r}}$ is a unit vector along the line joining the two charges and pointing away from q_2 toward q_1. This expression is Coulomb's law. The SI unit of charge is the *coulomb*, C, and the force is given in *newtons*, N. In SI units, the constant of proportionality k is usually given in terms of the permittivity of free space ε_0, as $1/4\pi\varepsilon_0$, with

$$\varepsilon_0 = 8.85 \times 10^{-12}\ \mathrm{C^2\,N^{-1}\,m^{-2}}.$$

The force is repulsive if the charges have equal signs and attractive if they have opposite signs.

doi:10.1088/2514-3433/ac878ech2

The force between point charges q_1 and q_2 is not affected by the presence of any other point charges in the vicinity. The total force on q_1 due to all the charges present is calculated by computing the force on q_1 due to q_2 alone, then calculating the force on q_1 due to q_3 alone, and so on, and adding all these contributions. This is known as the *principle of superposition*. Coulomb's law and the principle of superposition are the only two physics concepts required for electrostatics. The rest is mathematical manipulation of these concepts (Griffiths 2012).

2.2 The Electric Field

How is the force between two charges separated by a distance r transmitted from one charge to the other? The classical solution, introduced by Michael Faraday in the nineteenth century, involves the concept of *field*. According to Faraday, the space around an electric charge is distorted in such a way that any other charge placed in this space accelerates with a force that is given by Coulomb's law.

The electric field around a charge q can be mapped by placing a positive charge q_0 at different locations around q. The force at each one of these locations is given by Coulomb's law and points away from q (Figure 2.1). This force is

$$\mathbf{F} = \frac{q_0}{4\pi\varepsilon_0}\frac{q}{r^2}\hat{\mathbf{r}}$$

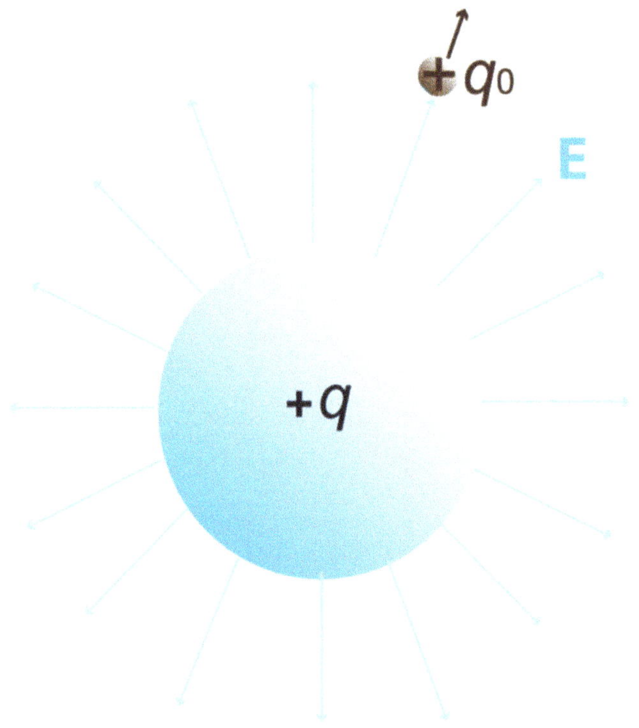

Figure 2.1. The electric field in the space around a positive charge q is mapped by placing a positive charge q_0 at different points around the charge q.

or

$$\mathbf{F} = q_0\mathbf{E}$$

where **E** is the electric field of the charge q:

$$\mathbf{E} = \frac{1}{4\pi\epsilon_0}\frac{q}{r^2}\hat{\mathbf{r}}. \tag{2.2}$$

2.3 Gauss's Law

The electric *field lines* around a charge q are the graphical representation of the electric field vectors **E** at different locations around the charge. From Equation (2.2), we can see that the field lines around a positive charge point radially outward from the charge and those around a negative charge point radially toward the charge. If we assume that the number of field lines is proportional to the magnitude of the charge q, we can see that if we place this charge inside a closed surface, all the field lines coming out (for a positive charge) or going in (for a negative charge) cross this closed surface. We can define the *flux* of **E**, ϕ, through any surface as the net number of field lines crossing this surface. Since the number of field lines is proportional to the magnitude of the charge, the flux is

$$\phi = \oint \mathbf{E} \cdot d\mathbf{a} = \frac{q}{\epsilon_0} \tag{2.3}$$

where $d\mathbf{a}$ is an element of area of the enclosing surface. Equation (2.3) is known as Gauss's law.

2.4 Electric Potential

If a charge q_0 moves from point 1 to point 2 in an electric field **E**, the work done by the field is

$$W = q_0 \int_1^2 \mathbf{E} \cdot d\mathbf{l}$$

where $d\mathbf{l}$ is an element of length along a path from 1 to 2. The work done by the field is independent of that path from point 1 to point 2 and depends only on these two points. This independence of the path taken makes the electric field *conservative*. The difference in the potential energy between the final point 2 and the initial point 1, as the electric field moves the charge q_0 from 1 to 2 is

$$\Delta U = U_2 - U_1 = -W.$$

The negative sign indicates that, as the electric field does work on the charge to move it between those two points, the potential energy of the point charge decreases, being less at point 2 than at point 1.

The potential energy of q_0 depends on the nature of the electric field **E** and on the magnitude of the charge. It is then convenient to define the potential energy per unit charge, V, which depends only on the nature of the field, as follows,

$$\Delta V = V_2 - V_1 = -\frac{W}{q_0}.$$

The quantity V is called the *electric potential.*

If we chose point 1 to be the standard reference point, which we can arbitrarily locate at infinity, and assign to the potential energy of q_0 when located at this point a value of zero, then V_1 is clearly zero and we can set V_2 to V. The potential at a point in an electric field is then,

$$V = -\int \mathbf{E} \cdot d\mathbf{l}. \tag{2.4}$$

2.5 Conductors in Electrostatic Fields

Conductors are a particular class of matter in which electric charges are free to move about *in* the material. In the case of solid metallic conductors, one or two electrons per atom are the charges that move about. For liquid conductors, such as salty water, it is ions that are free to move in the liquid. Gases at extremely high pressures, such as hydrogen deep in the atmospheres of Saturn and Jupiter, are also liquid conductors. In this case, it is electrons freed from their nuclei that are free to move.

Although conductors have free charges that can move about, an isolated conductor is still neutral. If an additional charge were to be placed inside an isolated conductor, the repulsive forces of the charges inside the conductor would push this additional charge out to the surface, where this mutual repulsion is balanced by surface forces. The time for this process to take place would be of the order of 10^{-14} s (Guru & Hiziroglu 1988). Thus, the net charge inside an isolated conductor is zero.

In electrostatics, there are no continuous sources of energy in a conductor and the electrons arrange themselves until the electric field everywhere inside the conductor is zero. This is true even if the conductor is placed in an external electric field. In this case, the external field exerts a force on the free electrons, moving them inside the conductor in an opposite direction to the field, leaving the positive ions in the metal unpaired. However, the conductor is still neutral. This separation of charges caused by the external field generates an internal electric field that opposes and cancels the external applied field, so that the net field inside remains at zero.

For the case of a charged spherical conductor, the charges are distributed uniformly over the surface of the sphere. The electric potential at the surface of a sphere of radius r holding a charge q is

$$V = Er = \frac{1}{4\pi\epsilon_0}\frac{q}{r}. \tag{2.5}$$

Using Gauss' law, it is not difficult to see that the electric field just outside a conductor is

$$E = \frac{\sigma}{\epsilon_0} \tag{2.6}$$

where σ is the surface charge density.

2.6 Capacitance

When charge is added to an isolated conductor, the charge stays on the surface of the conductor. The charge is now "stored" on the conductor. This charge can later be moved to another place. We can think of an isolated conductor as a place to store charge. If we have two isolated conductors nearby, we can use one as the source of charge and the other as the receptor by moving charges from one to the other. A separation of charges is produced, with one conductor holding with a charge $+q$ and the other a charge $-q$. This system is called a *capacitor*.

From Coulomb's law (2.1), we know that the electric field **E** that is set up between the two charged conductors is proportional to the charge q. The potential V between the two conductors is also proportional to q. The constant of proportionality is the *capacitance* C of the system of two conductors:

$$C = \frac{q}{V}. \tag{2.7}$$

Moving a charge (an electron) from one conductor to the other in a capacitor requires work. When the first electron is moved, a small electric field is set up across the two conductors. The electric field lines start at the positive ion left behind and end at the electron. Work must be done against this field to move the next electron across. This field clearly increases with the number of electrons transferred. If at a certain time a charge q has been transferred, the work needed to move the next dq is

$$dW = Vdq = \frac{q}{C}dq.$$

The work needed to charge a capacitor to a total Q is

$$W = \int_0^Q \frac{q}{C}dq = \frac{1}{2}\frac{Q^2}{C} = \frac{1}{2}CV^2.$$

The charged capacitor has a final potential V across the conductors.

2.7 Electrostatic Breakdown

The strength of the electric field around a conductor depends to some extent on the shape of the conductor. If the conductor has a sharp edge or point, for example, the field around the edge or point is much higher than on the other areas of the conductor.

Following Richard Feynman (Feynman et al. 1964), we can illustrate how the field is highest at the sharp edges by considering two charged conducting spheres of radii r and R (with r much smaller than R) connected by a thin wire (Figure 2.2). If the large sphere has a charge Q and the small sphere holds a charge q, the electric potentials at the surface of each sphere are (see Equation (2.5)):

$$V_R = \frac{1}{4\pi\epsilon_0}\frac{Q}{R} \quad \text{and} \quad V_r = \frac{1}{4\pi\epsilon_0}\frac{q}{r}.$$

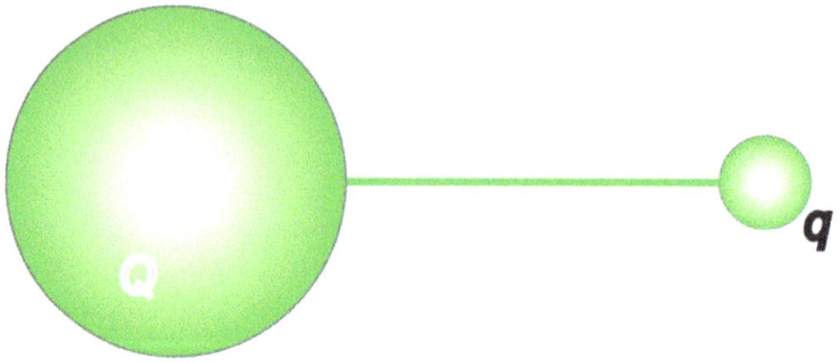

Figure 2.2. Two spheres of radii R and r holding charges Q and q, respectively, and connected by a thin wire can be used to illustrate how the field is highest at sharp edges.

But $V_R = V_r$, so that

$$\frac{Q}{R} = \frac{q}{r}.$$

Since the magnitude of the electric field at the surface of each sphere is proportional to the surface charge density (2.6),

$$\frac{E_R}{E_r} = \frac{\sigma_R/\epsilon_0}{\sigma_r/\epsilon_0} = \frac{Q/R^2}{q/r^2} = \frac{r}{R}.$$

Thus, the electric field is inversely proportional to the radius of curvature of the surface.

In real situations such as within an atmosphere, ions and electrons can be accelerated by the electric fields around conductors. If a conductor has sharp edges or points, the electric field can be large enough to accelerate the ions and electrons to the speeds needed to ionize molecules, generating additional ions and electrons that can produce an avalanche, leading to electrostatic breakdown in that atmosphere.

2.8 Dielectrics in Electric Fields

Dielectrics or insulators are materials in which all electrons are strongly bound to their atoms or molecules. Therefore, insulators have no conduction or free electrons. Since charge cannot flow freely in insulators, they have zero electrical conductivity. An external electric field **E** can distort the charge distribution in an insulator, causing the electrons around a molecule to be microscopically displaced. The displacement can cause the molecule to stretch, in the case of nonpolar molecules, or to rotate, in the case of polar molecules.

For a nonpolar molecule, the external field displaces the electrons away from the center of the molecule, in the opposite direction to the applied field **E**, and the nuclei of the atoms forming the molecule in the direction of the field. In this case, the centers of the positive charge and the center of the negative charge in the molecule are separated by a small distance, forming a small *dipole*. The molecule is said to be *polarized*.

When two opposite charges of equal magnitude q are separated by a distance s, it is convenient to define a vector quantity called the *dipole moment* **p** as

$$\mathbf{p} = q\,\mathbf{s}$$

where **s** is a vector of length s pointing away from the negative charge and toward the positive charge. For the case of an atom in an electric field **E**, the dipole moment can be written as

$$\mathbf{p} = \alpha\mathbf{E}$$

where α is the *atomic polarizability* of the atom.

In the case of polar molecules that have a permanent dipole moment built-in, such as the water molecule, the external electric field **E** will exert a torque on the molecule that will rotate the molecule to tend to align it with the field. For a molecule with a dipole moment **p**, the torque is

$$\boldsymbol{\tau} = \mathbf{p} \times \mathbf{E}.$$

The polarization of insulators with nonpolar molecules and the alignment of the molecules in insulators with polar molecules disappears almost immediately after the external field is removed. This is due to the continuous interactions of the molecules in a solid. These interactions increase with temperature.

The polarization of the insulator that results from the alignment of all the dipoles can be represented by the total dipole moment per unit volume, **P**. This quantity is known as the *polarization*. In linear dielectrics, the polarization is proportional to the field,

$$\mathbf{P} = \varepsilon_0\, \chi_e \mathbf{E}$$

where χ_e is the electric susceptibility of the medium.

When polarized by an external **E** field, an uncharged insulator remains neutral. Inside the insulator, the aligned dipoles form rows in which the positive end of a dipole is adjacent to the negative end of the next dipole. The opposite charges of equal magnitude from adjacent dipoles cancel each other out. Clearly the two ends of the row of dipoles are not paired up with opposite charges. The positive charge at the end is located near the insulator surface on the lee side of the field and the negative end on the side facing the field. These unpaired charges at the ends are called *bound* charges.

The build-up of opposite bound charges at each end of the dielectric by an external field **E** produces an internal electric field $\mathbf{E}_\mathrm{b}$ in the opposite direction to that of **E**. This is the field due to the polarization of the dielectric. The resultant field $\mathbf{E}_\mathrm{d}$ inside the dielectric is

$$\mathbf{E}_\mathrm{d} = \mathbf{E} + \mathbf{E}_\mathrm{b}.$$

$\mathbf{E}_\mathrm{d} < \mathbf{E}$, since **E** and $\mathbf{E}_\mathrm{b}$ are in opposite directions.

A non-uniform electric field produces different forces on each of the two opposite charges of a dipole. For an infinitesimal dipole, when s approaches zero, the force is given by

$$\mathbf{F}_\mathrm{dipole} = \mathbf{p} \cdot \nabla\mathbf{E}. \tag{2.8}$$

We can see from this expression that for the force on the dipole to have a non-zero value, the external electric field must be non-uniform.

2.9 Dielectrophoretic Forces

A non-uniform electric field will exert a force on an uncharged dielectric or conductive particle with an induced dipole moment. This *dielectrophoretic force* will also act on two or more particles with dipole moments induced by uniform or non-uniform external electric fields. In either case, only non-uniform fields generate dielectrophoretic forces.

Let's consider the case of a dielectric sphere of radius a and permittivity ε_p immersed in a dielectric fluid of permittivity ε_f in a region where there is an external uniform electric field of magnitude E_0. To obtain a good approximation to the expression for the dielectrophoretic force, we use the effective dipole moment concept. This is an ideal free-charge, point dipole that generates the same field as the real radius a particle when immersed in the same dielectric fluid. The expression for the effective dipole moment is

$$p_{\mathrm{eff}} = 4\pi\varepsilon_f K a^3 E_0 \tag{2.9}$$

where K is the Clausius–Mossotti function, given by

$$K = \frac{\varepsilon_p - \varepsilon_f}{\varepsilon_p + 2\varepsilon_f}. \tag{2.10}$$

The dielectrophoretic force can be written in terms of the effective dipole moment using Equation (2.8), or

$$\mathbf{F}_{\mathrm{DEP}} = \left(\mathbf{p}_{\mathbf{eff}} \cdot \nabla\right)\mathbf{E}. \tag{2.11}$$

Using Equations (2.9)–(2.11), the dielectrophoretic force becomes

$$\mathbf{F}_{\mathrm{DEP}} = 2\pi\varepsilon_f a^3 \left(\frac{\varepsilon_p - \varepsilon_f}{\varepsilon_p + 2\varepsilon_f}\right) \nabla E_0^2. \tag{2.12}$$

This well-known expression is quite general (Pohl 1978; Jones 1995). We note that the dielectrophoretic force depends on the value of the Clausius–Mossotti function K, which gives the strength of the effective polarization of the particle. When the permittivity of the particle is greater than that of the medium ($\varepsilon_p > \varepsilon_f$) the Clausius–Mossotti function is greater than zero. In this case, $\mathbf{F}_{\mathrm{DEP}}$ is positive and the particle is attracted to the maximum value of the electric field intensity. Conversely, when ($\varepsilon_p < \varepsilon_f$), $\mathbf{F}_{\mathrm{DEP}}$ is negative and the particle is attracted to the minimum value of the field intensity (Pohl 1978; Jones 1995). This principle underlies some of the dust mitigation technologies developed for planetary missions (see Chapter 7).

2.10 Plasmas

A highly ionized gas is a good conductor of electricity. The region in the gas with a balanced number of positive and negative charges is known as a *plasma*. When an ionized gas is placed in an external electric field, the ions and free electrons in the gas distribute themselves to shield most of the gas from the external field. The space charge at the boundary of the plasma is known as the *shield.*

If a small charge $+Q$ is placed in a plasma, electrons in the plasma are attracted to the charge while positive ions are repelled by the charge. The cloud of electrons surrounding the small positive charge Q shields the charge from other electrons, reducing their interaction. To an electron far away from Q, the electrons and positive ions balance out, so that the interaction with Q is the regular Coulomb interaction. An approaching high-energy ion can partially penetrate the cloud, leaving some of the shielding electrons behind. If the ion has an energy high enough to penetrate the cloud entirely, it will feel the full Coulomb repulsion.

In equilibrium, the electron density is

$$N_e = N_0 \mathrm{e}^{\frac{U-U_0}{kT}}$$

where U is the potential of the electron in the field of the charge Q, U_0 is the plasma potential, N_0 is the electron density where $U = U_0$, T is the absolute temperature, and k is Boltzmann's constant (Reitz et al. 2008). A similar expression holds for the ion density. The potential U of a charge at a distance r from Q is given by

$$U = U_0 + \frac{Q}{4\pi\epsilon_0}\mathrm{e}^{-\frac{r}{h}}$$

where h is the *Debye length* or shielding distance, given by

$$h = \sqrt{\frac{\epsilon_0 kT}{2N_0 e^2}}.$$

Plasmas are common in the solar system, particularly at airless bodies. Plasmas can also be found in the regions surrounding planetary atmospheres. In these regions, there are relatively few neutral particles, and ionization from solar ultraviolet radiation and energetic particles produces ions and electrons in sufficient quantities for a plasma to exist. Many plasmas in space contain small particles in addition to the plasma; these environments are known as *dusty plasmas* (Merlino & Goree 2004). Some of the planetary environments considered elsewhere in this book, such as the surface of the Moon (Chapter 7), or an asteroid (Chapter 8) are similar to a dusty plasma. In planetary atmospheres the presence of many neutral gas molecules absorbs some of the ionizing energy and leads to a lower fraction of ionized particles. Atmospheres are typically only weakly conductive with respect to the plasma space environment around them.

References

Feynman, R. P., Leighton, R. B., Sands, M., et al. 1964, Feynman Lectures on Physics, Vol. 2 (Reading, MA: Addison-Wesley)

Griffiths, D. J. 2012, Introduction to Electrodynamics (4th ed.; London: Pearson Education)

Guru, B. S., & Hiziroglu, H. R. 1988, Electromagnetic Field Theory Fundamentals (2nd ed.; Cambridge: Cambridge Univ. Press)

Jones, T. B. 1995, Electromechanics of Particles (Cambridge: Cambridge Univ. Press)
Merlino, R., & Goree, J. 2004, PhT, 57, 3
Pohl, H. A. 1978, Dielectrophoresis (Cambridge: Cambridge Univ. Press)
Reitz, J. R., Milford, M. J., Christy, R. W., et al. 2008, Foundations of Electromagnetic Theory (4th ed.; Reading, MA: Addison-Wesley)

Carlos I Calle and Karen L Aplin

Chapter 3

Space Radiation Fundamentals

3.1 Sources and Types of Radiation Found in Space

Space contains a mixture of many different types of radiation from many sources. Radiation can be particles, for example alpha particles, or electromagnetic waves, such as ultraviolet or gamma radiation. The theory of wave–particle duality demonstrates that electromagnetic waves can behave as particles under certain circumstances. However, all the principles described in this chapter can be adequately explained by treating the entire electromagnetic spectrum as being made up of waves.

Space is full of highly energetic particles called *cosmic rays*, which are generated by powerful astrophysical events such as supernovae. Most cosmic rays are atomic nuclei, mainly hydrogen nuclei (protons), helium nuclei (alpha particles), and about 1% are the nuclei of heavier elements. They have a wide range of energies extending from 10^9 up to 10^{21} eV (Hillas 1972). (The energy of individual particles is described in terms of electron volts, eV, where 1 eV is 1.6×10^{-19} J.) The most energetic cosmic rays have been described as having the energy of a cricket ball or baseball traveling at 160 km h^{-1} (Friedlander 2012).

Our Sun's surface is at about 6000 K. It emits electromagnetic radiation in the wavelength range from 250 to 2500 nm, similar to that of a blackbody emitter. The thermal radiation emitted by all blackbodies at the same temperature has an identical spectrum. The power radiated per unit area by a blackbody increases rapidly with temperature. In 1879, the Austrian physicist Joseph Stefan found empirically that the power per unit area R is proportional to the fourth power of the temperature T,

$$R = \sigma T^4$$

where

$$\sigma = 5.67 \times 10^{-8}\ \mathrm{W\,m^{-2} \cdot K^{-4}}$$

doi:10.1088/2514-3433/ac878ech3

is the Stefan–Boltzmann constant. This equation is known as Stefan's law. Figure 3.1 shows the spectral distribution of a blackbody as a function of wavelength λ for different temperatures.

The peak wavelength emitted depends on the temperature, as seen in Figure 3.1. What is interesting in those curves is that the peak wavelength is inversely proportional to the temperature, or

$$\lambda_{\mathrm{max}} \propto 1/T.$$

This expression is known as Wien's displacement law, after the German physicist Wilhelm Wien. We can see that the hotter the object emitting radiation, the shorter the wavelength at which this radiation is emitted. The peak emission of the Sun at the surface temperature of 6000 K is 483 nm.

The Sun's activity varies on a range of timescales. It has an underlying cycle of 11 years, over which its total irradiance varies slightly, and its activity varies almost on a day-to-day basis. The solar cycle has been known about since the 18th century because of long-term observations of sunspots, which are locally cooler regions on the surface of the Sun associated with magnetic activity (Figure 3.2). Modern observations show that in addition to sunspot variations, the Sun emits bursts of particles and radiation, which can occur at any time but are more likely at the maximum of the 11 year cycle. This day-to-day variability is called *space weather*, by analogy with terrestrial weather. Space weather events are of increasing concern as society becomes more dependent on space-based technologies and complex electronics systems, which are susceptible to particles of solar origin, as will be described in Section 3.3. The magnetic and electromagnetic influence of the Sun is called the *heliosphere* and extends out to beyond Pluto. The entire solar system is therefore

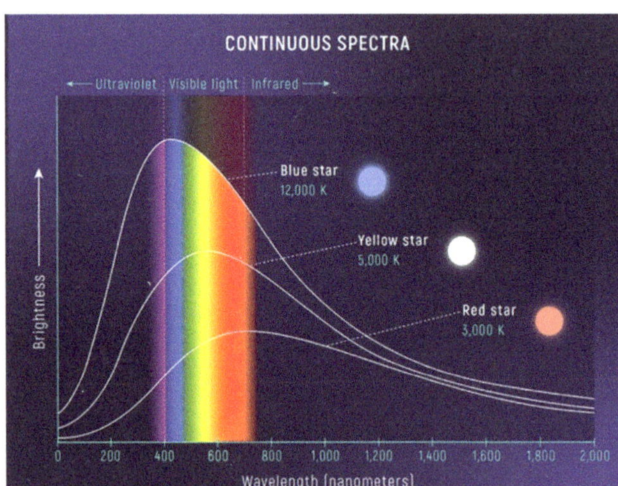

Figure 3.1. The spectral radiance (power per unit area) R of a blackbody as a function of wavelength λ at different temperatures. (Courtesy NASA, Space Telescope Science Institute.) The Sun is a typical yellow star.

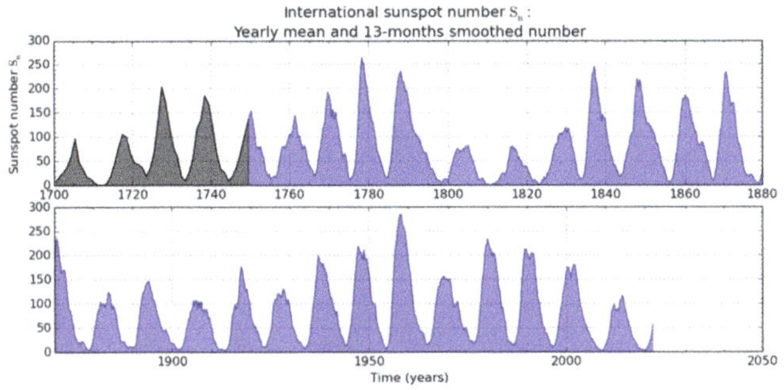

Figure 3.2. Annual mean sunspot number from 1700 to 2022. The region shown in dark gray, before 1748, was when the Sun was not observed daily. For this period the averages are only computed for those days that observations were made. (Courtesy of SILSO, Royal Observatory of Belgium, Brussels.)

influenced by solar radiation, but space weather events are only considered to be relevant for the terrestrial planets relatively close to the Sun.

Atmospheres are ionized by cosmic rays, which leads to electrostatic effects discussed elsewhere in this book. In addition to cosmic rays, atmospheres can become ionized by natural radioactivity emitted from rocky surfaces, such as terrestrial land masses (Aplin 2013) in the form of alpha and beta particles and gamma radiation. Some atmospheres, such as that of Mars—which lacks the protective ozone layer of Earth—are ionized by ultraviolet radiation from the Sun. Planetary moons with atmospheres are often affected by radiation from their host star, such as Saturn's largest moon, Titan, the upper layers of which can be ionized by particles trapped in Saturn's magnetosphere (Aplin 2013).

3.2 Measurement of Energetic Particles in Space

There are many approaches to measure energetic particles in space. Miniaturized versions of detectors such as Geiger counters, which produce a count for each energetic particle, are one of the simplest methods. Scintillators, which release a pulse of light for every energetic particle, are also used and in many cases can be more helpful than Geigers since they allow determination of the energy of the particle from the brightness of the light emitted. These simple instruments were some of the first to be flown in space. The first successful US spacecraft, Explorer 1, (launched 1958 February 1) carried Geiger counters and scintillators and was the first to detect the trapped radiation around the Earth known as the Van Allen belts (Ludwig 1959; see Chapter 5).

As radiation instrumentation is commonly carried on spacecraft, there have been many measurements of radiation in space. Particle detectors were regular payloads during the early days of space exploration in the 1950s and 1960s. As technology has progressed, energetic particle instrumentation has become more sophisticated and can now provide detailed information, such as the results from the instrument named RAD on NASA's Curiosity rover on Mars (Hassler et al. 2014). The Mars surface environment is subject to higher radiation levels than Earth, even though the cosmic ray and space weather particles arriving at each planet are similar. This is

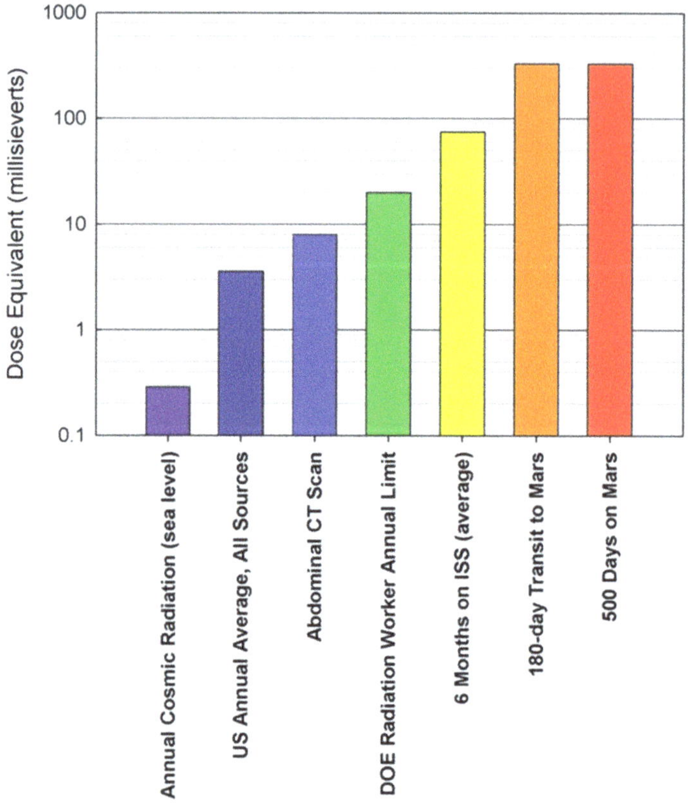

Figure 3.3. Comparison of the radiation dose equivalent for a 500 day surface stay to that from a 180 day transit to Mars based on the Curiosity rover data, a 6 month stay on the International Space Station (ISS) at 420 km altitude during the 2013 solar maximum, and several Earth-based sources of radiation. Reproduced from Hassler et al. (2014), with permission from AAAS.

because the thin atmosphere of Mars is not able to absorb as much radiation as Earth's thicker atmosphere. A return journey to Mars with a stay of less than two Earth years would involve an astronaut being exposed to over 100 times more radiation than on Earth, Figure 3.3. Radiation protection is therefore a key consideration for future human space exploration and will be discussed in Sections 3.3 and 3.4.

In common with many planetary radiation payloads the RAD instrument was also able to determine the energy spectrum of incoming particles and separate radiation from cosmic rays and space weather events. Space weather events can bring energetic protons down to the Martian surface. In its first eight years of operation, RAD detected five peaks in the dose rate from space weather. The total dose from a 2 h medium-sized space weather event on Mars is 8 mSv, equivalent to a chest CT scan (Guo et al. 2021). The SI unit Sv is the Sievert, a unit of ionizing radiation that measures its effect on the human body.

At Earth the main problem caused by space weather is not the radiation dose to humans on the surface but possible technological issues, discussed in more detail in Section 3.3. Society's increased reliance on electrical power distribution networks, and, increasingly, satellite technology, has motivated the operational monitoring and forecasting of space weather events based on data from several Earth-orbiting satellites. These can be in low-Earth orbit, responding to fluctuations in near-Earth activity, or in geostationary orbit at 32,500 km where they are able to give earlier warning of space weather events. This capability is helpful as space weather events can be highly complex, difficult to predict, and occasionally damaging. More information on the near-Earth electrostatic environment is provided in Chapter 5.

3.3 Effects of Space Radiation

Radiation interacts with any matter that it encounters. Roughly speaking, the most high-energy particles will pass rapidly through matter losing almost no energy *en route*, and lower-energy particles deposit all their energy in matter. Some particle interactions with matter produce new radiation called *secondary radiation*. This leads to a range of effects depending on the type of particle and its energy, and the type of matter involved.

One of the major effects of radiation on matter is to ionize it. As was mentioned in Section 3.1 earlier, atmospheres are ionized by cosmic rays and other energetic particles and waves. Ionization means the molecules are broken up and charge is released, and this can occur in any matter, from atmospheric gases to solids, like semiconductors. Non-ionizing radiation, such as neutrons, also interacts with the nuclei of atoms in matter (Krane 1987). Energetic particles in space are mainly ionizing, but cosmic rays can create neutrons as secondary particles when they interact with matter. Both ionizing and non-ionizing radiation can cause problems for electronic circuits and biological matter, such as humans and animals.

Biological matter can be damaged by space radiation which causes mutations in DNA, leading to an increased risk of cancer in the long term. These risks are identical to those experienced by workers on Earth who receive higher radiation doses than background, such as flight crew. Extremely high radiation doses can also cause radiation sickness, though these levels of radiation are highly unlikely to occur from natural radiation in space, given that the dose from space weather storms measured by the RAD instrument on Mars (and discussed in Section 3.2 above) is comparable to a CT scan. Just like other workers exposed to radiation, upper limits are specified for space travelers by their employers, the space agencies, which are in many cases higher than the limits for staff at nuclear facilities (Straube et al. 2010).

Electronic systems can be disturbed by radiation in several ways. One-off ionizing events, when a high-energy particle hits the depletion region of a semiconductor, are called *single event effects* (or *single event upsets*), which can cause errors by switching a 0 to a 1 or vice versa, or a hardware short circuit, called a *latch up*. In the worst case, single event effects can cause catastrophic failure of the component. There is no clear definition of "high-energy" particles that create single event effects, but it is assumed to be 1–50 MeV (1 MeV = 1 mega (million) electron volts). Energetic particles can cause harmful electrostatic effects on and in spacecraft; these are discussed in Chapter 6. Longer-term exposure from lower-energy ionizing radiation can also lead to damage in semiconductor structures by continued degradation.

Satellites orbiting near Earth are more susceptible to the single event problems caused by an individual particle, whereas spacecraft in interplanetary orbits for many years are more vulnerable to the cumulative dose effects. Space weather events are a significant hazard to Earth-orbiting satellites and there is considerable international effort to evaluate and forecast these hazards. Missions to the Sun and inner planets become increasingly more susceptible to space weather effects the closer they get to the Sun. In terms of interplanetary space travel, the radiation environment of Jupiter is challenging for spacecraft due to its particularly energetic and large radiation belts, in which particles from the solar wind are accelerated by Jupiter's strong magnetic field (see Chapter 11). Approaches taken to mitigate damage from radiation in space are outlined in the next section.

3.4 Mitigating Radiation Effects

Humans living on the Earth's surface are protected from space radiation by the atmosphere. The magnetic fields of both the Earth and Sun provide further protection for Earth and its near-space environment (see Chapter 5 for more information). As the Sun's magnetic field is stronger during the most active part of its 11 year cycle, this is when the cosmic ray flux is at a minimum. The risk to spacecraft and human explorers in the near-Sun heliosphere is enhanced at solar maximum due to the increased likelihood of space weather events, even though the cosmic ray flux is lower. Radiation protection is therefore a key part of planning both crewed and uncrewed space missions, to maximize their chances of success.

Careful choice of orbit can help, for example, the Van Allen belts (which are discussed in more detail in Chapter 5) are normally avoided by low-Earth orbiting satellites. Jupiter mission trajectories are planned to balance the requirements to minimize both the mass of fuel needed and the radiation dose to the spacecraft (Kayali et al. 2012). Electrical circuits can be designed with *radiation hardened* chips that are tolerant of a high-radiation environment. However, these can be expensive so are not always appropriate for lower-cost or shorter-duration missions. Design choices can also be included in the circuits, such as latch-up protection, to mitigate against single event effects.

The shielding of spacecraft and, potentially, human planetary habitats is an important aspect of preventing against radiation damage. Dense materials such as lead and gold effectively absorb ionizing radiation with energies up to hundreds of MeV, with thicker layers absorbing higher energy particles. For space applications mass is critical, so careful calculations are required, usually ruling out the use of thick layers of dense material. Thin layers of less dense metals such as aluminum can screen out a reasonable fraction of ionizing radiation. Figure 3.4 shows that most of the dose reduction occurs by adding a few millimeters of aluminum shielding with proportionately little benefit of adding extra shielding. The absorbed radiation dose (with units called rad) is the amount of energy from any type of ionizing radiation

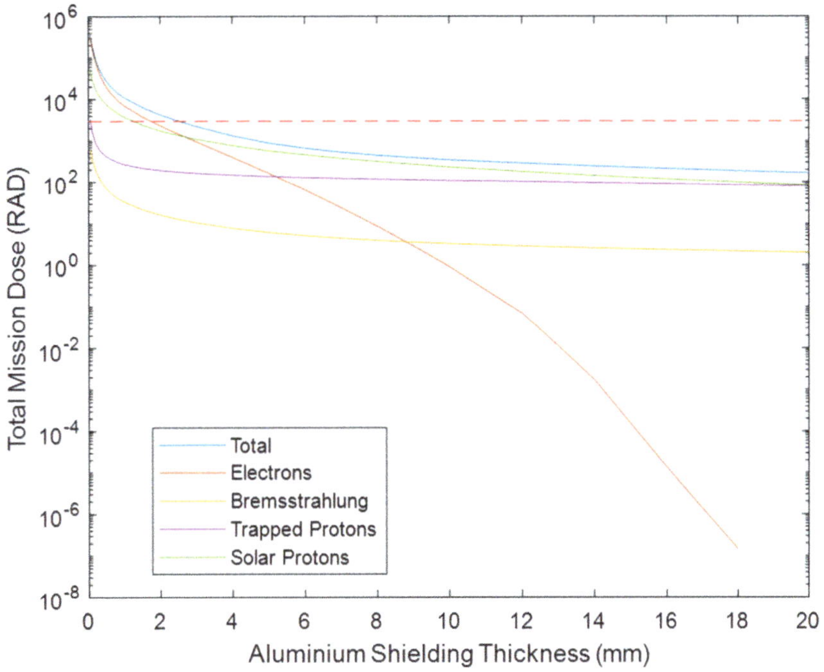

Figure 3.4. Effect of shielding a small spacecraft in low-Earth orbit (500 km) with aluminum. The total ionizing dose in rads over a mission of 1.5 years from a range of space-weather related particles is plotted against the shielding thickness. The total dose is not significantly improved by adding more than about 5 mm of aluminum. The red dashed line indicates the anticipated failure dose for a small satellite. Reproduced from Timperley & Berthoud (2022) with permission from the authors.

deposited in a medium, such as human tissue. This shielding can often be achieved by the material of the spacecraft itself. Shielding against the energetic particles responsible for single event effects is difficult to achieve since they would require a much thicker layer of aluminum or a denser material. Since the risk from a single event effect is essentially random, single event effects are not often mitigated against in Earth-orbiting spacecraft.

For the especially difficult environment experienced by Jupiter-orbiting satellites, extra precautions are required, such as housing the electronics in a shielded vault. The NASA Juno mission to Jupiter had a titanium vault with the most radiation-sensitive instruments placed in the center of the vault to be shielded by the other instruments in addition to the titanium walls (Kayali et al. 2012). Recent designs use optimization techniques, suggesting, for example, that inclusion of thin lead and magnesium layers would reduce the mass needed by about 40% compared to aluminum screening (Wang et al. 2019).

There is significant research into how to protect humans from radiation in space. Travel to Mars requires special thought, as the crew will experience long-term radiation exposure as a combination of their journey in the near-Sun environment, and when based on the Martian surface. “Storm shelters”, areas of extra radiation screening, within space stations are planned where the crew can wait during a space weather event. These could be similar to the vaults discussed above; another option is to create a magnetic field to deflect particles away (Bamford et al. 2014). On the Martian surface, use of materials that are naturally present to remove the need to carry shielding from Earth is preferred. Some studies suggest that around 1 m of Martian regolith (soil) would shield a significant fraction of the radiation experienced at the surface, so shallow underground habitats could be built (Llamas et al. 2022). A similar approach would be to exploit Mars' patchy magnetic field to create so-called "mini-magnetospheres" where explorers will be protected from radiation (Bamford et al. 2014). Human ingenuity will need to be fully deployed to protect future travelers to Mars and beyond from space radiation.

References

Aplin, K. L. 2013, Electrifying Atmospheres: Charging, Ionisation and Lightning in the Solar System and Beyond (Dordrecht: Springer Netherlands)

Bamford, R. A., Kellett, B., Bradford, J., et al. 2014, AcAau, 105, 385

Friedlander, M. 2012, Natur, 483, 400

Guo, J., Zeitlin, C., Wimmer-Schweingruber,, R. F., et al. 2021, A&ARv, 29, 8

Hassler, D. M., Zeitlin, C., Wimmer-Schweingruber,, R. F., et al. 2014, Sci, 343, 6169

Hillas, A. M. 1972, Cosmic Rays (Oxford: Pergamon)

Kayali, S., McAlpine, W., Becker, H., & Scheick, L. 2012, Juno radiation design and implementation, IEEE Aerospace Conf. 1

Krane, K. S. 1987, Introductory Nuclear Physics (Chichester: Wiley)

Llamas, H. J., Aplin, K. L., & Berthoud, L. 2022, P&SS, 218, 105517

Ludwig, G. H. 1959, RScI, 30, 223

Straube, U., Berger, T., Reitz, G., et al. 2010, AcAau, 66, 963

Timperley, L., & Berthoud, L. 2022, Reliability Analysis and Failure Mitigation Strategies for the PROVE Pathfinder CubeSat Payload, Proc. IEEE Aerospace Conf.

Wang, J.-Z., Ma, J.-N., Qiu, J.-W., et al. 2019, ITNS, 66, 2179

Chapter 4

Electrical Breakdown and Charge Decay in Planetary Atmospheres

For the most part, electrostatic phenomena are determined by the physical and chemical properties of the different planetary atmospheres. The maximum amount of electrical charge that can be deposited on surfaces, the length of time it takes for these surfaces to discharge, the nature of this discharge, and other factors, depend on atmospheric properties.

In general, planetary atmospheres can be considered to be spherically symmetric. This is actually an approximation, since sunlight generates day–night and latitudinal modifications to an atmosphere (Chamberlain & Hunten 1987). An atmosphere has a vertical structure that depends mainly on its atmospheric pressure and temperature.

Atmospheric pressure is defined as the normal force exerted by atmospheric gas per unit area. This pressure changes with elevation. At any given height, the force is the weight of the atmospheric gas above that height. Atmospheric temperature changes in a more complex way. In the case of the terrestrial atmosphere, the temperature decreases with altitude in some regions of the atmosphere but increases with increasing altitude in other regions.

4.1 Electrical Breakdown in Planetary Atmospheres

The strength of electric fields and the behavior of electrical charges either in the atmosphere or on the surfaces of materials depends mostly on the atmospheric properties that we have just discussed. The amount of charge that can accumulate on the surface of a material depends on the atmospheric pressure and electrical conductivity, the physical properties of the material, and on the distance to other materials. Whether the charge remains on the surface or is transferred to nearby

doi:10.1088/2514-3433/ac878ech4

objects depends on the values of these parameters. When charge does transfer across a gap, electrical breakdown occurs.

John S. Townsend and collaborators studied the phenomenon of electrical breakdown in depth in the 1940s (Townsend 1947). According to Townsend, if an electric field E is set up between two metal electrodes in a gas, any ions present in the gap between the electrodes will be accelerated in the field, ionizing some gas molecules and releasing electrons. In this case, N_0 electrons with mobility b will drift with a velocity u along the field lines. These electrons will collide with atoms in the gas. Depending on their kinetic energies, these electrons will ionize α atoms, releasing an equal number of electrons per unit path length, and will attach to η gas atoms per unit length. The coefficients α and η are the primary ionization and attachment coefficients. There are also secondary ionization processes, which include the release of electrons from the cathode by ion bombardment, the ionization of gas by positive ions and by photons, and photoemission by photons produced in the gas by electron ionization. A secondary ionization coefficient γ combines all these secondary processes. These three coefficients are known as the *Townsend coefficients* and are fairly well determined for many common gases.

If the potential across the electrodes is not too high, resulting in a relatively weak electric field, the number of electron–ion pairs that are formed decreases and the electrons and ions recombine. If the field is increased, the number of charges increases and, at some point, there is a transition to one form of sustained discharge. This transition constitutes a breakdown in the gas, called a spark, and usually occurs suddenly. For the spark discharge to be self-sustaining, there must be one secondary electron for every initial electron; that is, for a path length dl,

$$\gamma e^{\alpha-\eta}\mathrm{d}l - 1 = 1. \tag{4.1}$$

The Townsend coefficients α, η, γ depend on the local electric field. This field is a function of the space charge density and is not uniform (Cross 1987).

The breakdown potential depends on the product of the gap length d between electrodes and the gas number density. The relationship is known as Paschen's law (Paschen 1889). Historically, however, instead of the gas molecular number density, the pressure p of the gas is used in the expressions for Paschen's law. Figure 4.1 shows experimental values of the Paschen spark–breakdown potential as a function of pd for air and carbon dioxide.

Clearly, large values of pd are due to either higher pressures or large gaps. At higher pressures, there are enough molecules per unit volume for electron collisions with these molecules to be frequent, requiring larger potentials to accelerate the electrons so that they acquire sufficient kinetic energy for ionization to occur. Similarly, if the gap is large, there are enough molecules per unit volume even at moderate pressures for electron–molecule collisions to be frequent, thus requiring larger potentials for ionization. As either p or d decreases, the voltage required for ionization decreases to a minimum value.

At very low pressures, the gas density is low and collisions are rare. Therefore, larger potentials are possible before breakdown takes place. Likewise, with very

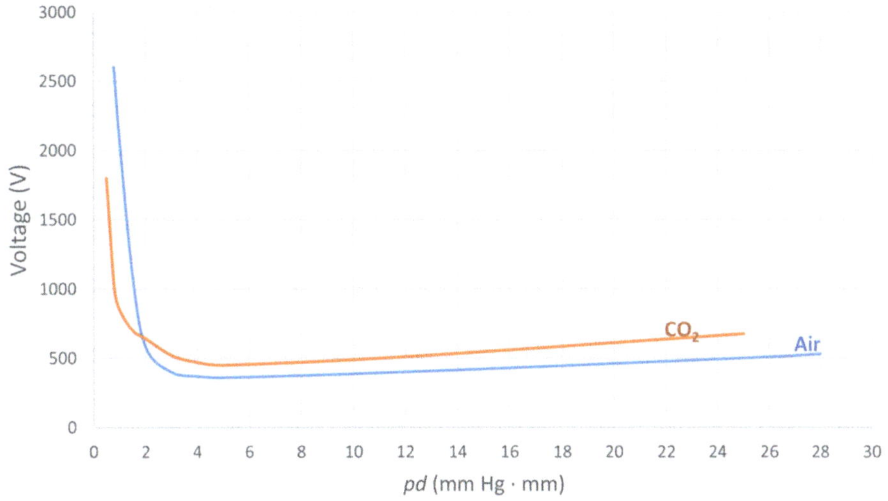

Figure 4.1. Paschen breakdown potential as a function of pressure multiplied by distance for air and carbon dioxide.

small gaps, electrons are not able to reach the kinetic energies needed for ionization, and breakdown takes place at much higher potentials.

The data shown in Figure 4.2 were obtained with clean, parallel plate electrodes without edge effects, so the field in the region between the electrodes at any point is never higher than at the center of the plane.

Electrostatic breakdown occurs naturally in planetary atmospheres. One important example is lightning. Although the processes that lead to lightning on Earth are not completely understood, the overall main process results from a large electrostatic potential difference between clouds and the ground, or between two clouds, that exceeds the local breakdown threshold of the atmosphere. Some types of lightning

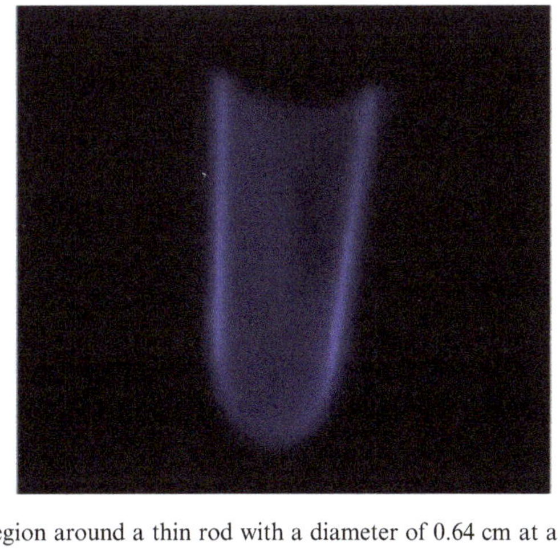

Figure 4.2. Corona in a region around a thin rod with a diameter of 0.64 cm at a 2.2 kV potential difference with a cylinder 9.6 cm in diameter. The corona current was 200 μA. (Courtesy of J. S. Clements, NASA.)

can also move upwards from a cloud to the ionosphere, the conductive upper layer of the atmosphere. These *transient luminous events* (TLEs) are a family of different types of upward discharge with evocative names such as sprites, elves, and blue jets. In addition to Earth, lightning occurs on Jupiter, Saturn, Uranus, and Neptune, and may occur on Venus. It is also anticipated to occur on Mars, but has not yet been detected (Aplin 2022).

4.2 Glow Discharges and Ion Wind

A glow discharge, also known as a *corona* discharge, takes place at a lower voltage than a spark discharge. When a high voltage is set up between two electrodes, one of which has a radius of curvature that is much smaller than the gap, a strongly non-uniform electric field is formed in the region. In the vicinity of the sharp electrode, the electric field is very high. A free electron in this region will accelerate, knocking off electrons from neutral atoms and creating electron–ion pairs. The newly freed electron is also accelerated in the strong field, producing additional electron–ion pairs. In this way, an electron avalanche is formed in the region close to the sharp electrode. The ionization process emits electromagnetic radiation. In the atmosphere of the Earth, this radiation is in the blue region of the electromagnetic spectrum. The shape of this glow is in the form of a tall crown, giving it its name of *corona* discharge (Figure 4.2).

If the voltage between the electrodes is not increased, any electrons in regions farther away from the sharp electrode do not gain enough kinetic energy to ionize the gas atoms and the discharge remains localized and stable. There is no breakdown across the electrodes.

When the sharp electrode is negative (*negative corona*), the electrons outside the high-field region, without enough energy to release other electrons from neutral atoms, attach themselves to some of these atoms, forming negative ions. These ions travel at low speeds toward the positive electrode and constitute the only current outside the high-field region. There is no glow in this external region. The positive ions formed by electron collisions with neutral atoms in the high-field region accelerate toward the sharp, negative electrode, where they release secondary electrons that contribute to the discharge mechanism, as happens in the electrical breakdown case.

When the sharp electrode is positive (*positive corona*), the secondary mechanism is less important. The electrons created by ionization in the high-field region accelerate toward the positive sharp electrode while the positive ions move toward the negative electrode.

In both cases, the slow-moving ions outside the high-field region, which have the polarity of the sharp electrode, collide with the atmospheric gas molecules, setting them in motion. The gas molecules move away from the sharp electrode, forming what is known as an *ion wind*. In the Earth's atmosphere, the ion wind speeds are about 1 m s^{-1}.

The voltage at which a stable corona is established depends on the gas and the electrode geometry. With very small and smooth electrodes, slowly increasing the voltage above the corona threshold results in an increase in the corona current that is proportional to the square of the applied voltage. This enhanced corona is called an *ultracorona*.

With larger electrodes, an increase in the corona voltage breaks up the corona glow into separate filaments or streamers (Figure 4.3). This *streamer corona* is still a stable glow discharge, with the current being carried by the ion flux. Increasing the voltage further will increase the length of the streamers until they reach the opposite electrode, at which point breakdown will occur.

Corona discharges occur naturally on Earth, normally near thunderstorms, where they are commonly called St. Elmo's Fire. Corona discharge may occur on many other planets, such as Mars, Jupiter, Saturn, and Venus.

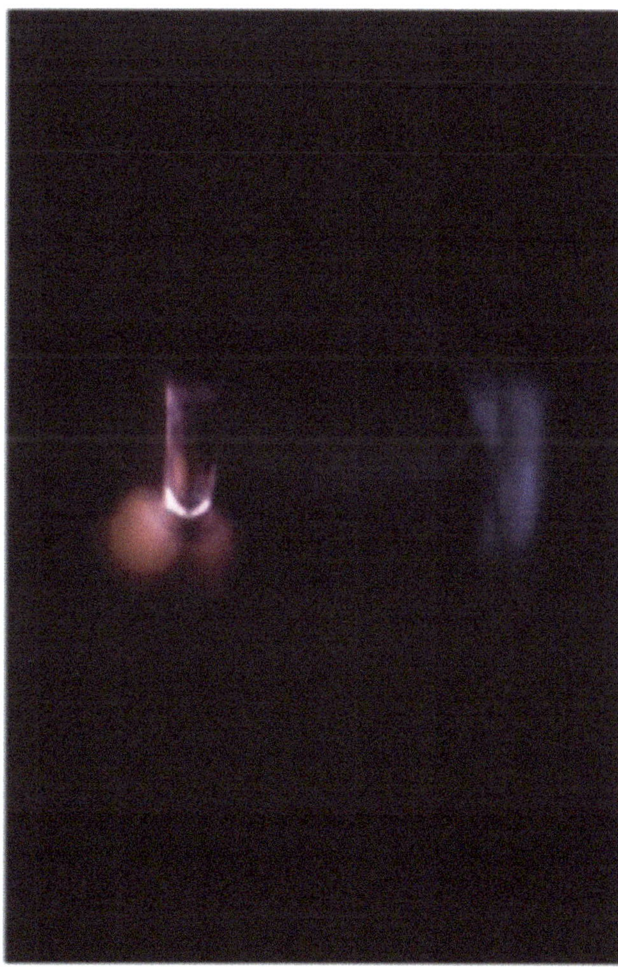

Figure 4.3. The pink glow of a streamer corona in air. (Courtesy of J. S. Clements, NASA.)

4.3 Charge Mobility

Electrical breakdown and glow discharges in planetary atmospheres are affected by the conductivity of the atmosphere, the presence of ions, and the ability of those ions and of charged particles to move. *Charge mobility* is a measure of the ability of an electric charge to move in an atmosphere while in the presence of an electric field.

The equation of motion for a charged particle in an external electric field E is

$$F = \frac{dp}{dt} = qE$$

where p is the particle's momentum, q is the particle's charge, and F is the electric force. If the particle moves with an average velocity v, the force imparted on the particle during a time t_c between collisions, the equation of motion can be written as

$$\frac{mv}{t_c} = qE$$

where m is the particle's mass. The average distance between collisions is called the *mean free path*. The particle's average velocity is then

$$v = \frac{qt_cE}{m} = \mu E \tag{4.2}$$

where μ, written as

$$\mu = \frac{qt_c}{m} \tag{4.3}$$

is the *charge mobility* of the particle. If the particle has a fixed charge and mass, such as an electron of charge e and mass m_e, and moves in an atmosphere with a given mean free path, the charge mobility is constant. Thus the average electron velocity is proportional to the electric field and the constant of proportionality is the electron charge mobility.

The electron mean free path in the Earth's atmosphere near the surface is 6.63×10^{-8} m (National Oceanic and Atmospheric Administration National Aeronautics and Space Administration and U S Air Force 1976). On Mars, with an atmospheric pressure about one hundredth of the Earth's, the mean free path near the surface is roughly 100 times longer or 4.8×10^{-6} m. The consequence of this is that free electrons are not normally found in Earth's near-surface atmosphere as they attach to atmospheric gases and form ions almost immediately.

4.4 Charge Decay and Conductivity in Planetary Atmospheres

Electric charges at rest on the surface of an isolated object will remain on that surface for a length of time that is mostly determined by the physical properties of the atmosphere in which the object resides. Charges need a conductive path to leave the surface. This conductive path is usually provided by ions present in the atmosphere. Cosmic rays and similar types of ionizing radiation reaching the atmosphere in the vicinity of the object generate ions as they interact with the neutral molecules. In the absence of an electric field, these ions are distributed more or less uniformly in the volume around the object.

The process generates positive and negative ions. When charge is present on the surface of the object, the electric field produced by this charge accelerates these atmospheric ions, thus generating an electric current in the atmosphere.

If the charged object is a conductor with surface charge density σ, the electric field E around the conductor is

$$E = \frac{\sigma}{\epsilon_0}. \tag{4.4}$$

The ions present in this electric field are accelerated by the field, generating an electric current j given by

$$j = \frac{E}{\rho} \tag{4.5}$$

where ρ is the electrical resistivity of the atmosphere in this region. Both positive and negative ion currents will be present, with the current of the same polarity as that of the charge on the conductor flowing away from the conductor and the current of opposite polarity flowing toward the conductor (Figure 4.4). The ion current flowing toward the conductor will clearly cause the charge density of the conductor to decrease at a rate $d\sigma/dt$, given by Equations (4.4) and (4.5):

$$j = \frac{d\sigma}{dt} = \frac{1}{\epsilon_0 \rho}\sigma. \tag{4.6}$$

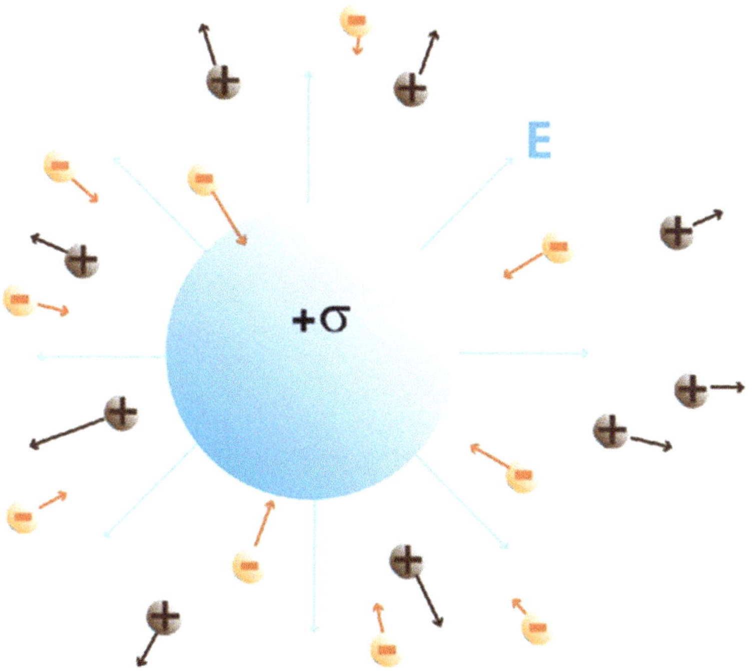

Figure 4.4. Ion currents in the vicinity of a charged conductor.

Since ρ is constant in the vicinity of the conductor, the solution to this differential equation is

$$\sigma = \sigma_0 \, e^{-t/\varepsilon_0 \rho}$$

or

$$\sigma = \sigma_0 \, e^{-t/\tau} \qquad (4.7)$$

where $\tau = \varepsilon_0 \rho$ is the *time constant* and σ_0 is the initial charge density of the conductor. As Equation (4.7) indicates, charge decay in a conductor is exponential. The time constant τ is the time it takes the charge to decay to $1/e$ of its initial value. The time constant defines the *electrical conductivity* of air, which depends on the number of ions (and/or electrons) present and their mobilities (Aplin & Harrison 2000).

Charges on insulators are not free to move and the mechanism of charge decay is different. Depending on the nature of the insulator, the condition of its surface and of any gases adsorbed, as well as on the conductivity of the atmosphere, charge may move through the insulator and/or across its surface (Kinderberger & Lederle 2008). Surface charge on an insulator can also be neutralized by ions in the atmosphere. As was the case with the charged conductor, an electric field arises due to the surface charge on the insulator. Because the charge on the insulator is not necessarily uniformly distributed around the surface, the electric field lines are not always perpendicular to the surface. At the edges of a charged patch on the surface, the field lines will point away from the normal and may even bend to fall on an area of the surface that may have a charge of opposite polarity. Ion currents will flow along those field lines bringing ions of opposite polarity that will neutralize some of the charge on the insulator surface. Charge decay in insulators does not follow the exponential decay form of Equation (4.7). This more complex decay follows a modified hyperbolic law (Seaver 2002).

References

Aplin, K. L. 2022, A&G, 63, 4.12

Aplin, K. L., & Harrison, R. G. 2000, RScI, 71, 8

Chamberlain, J. W., & Hunten, D. M. 1987, Theory of Planetary Atmospheres (San Diego, CA: Academic), 3

Cross, J. 1987, Electrostatics: Principles and Applications (Bristol: Adam Higler)

Kinderberger, J., & Lederle, C. 2008, ITDEI, 15, 949

National Oceanic and Atmospheric Administration National Aeronautics and Space Administration and U S Air Force 1976, US Standard Atmosphere (Washington, DC: US Government Printing Office)

Paschen, F. 1889, AnP, 37, 69

Seaver, A. 2002, ESA-IEJ Joint Meeting (Chicago, IL: Northwestern Univ.)

Townsend, J. S. 1947, Electricity in Gases (Cambridge: Cambridge Univ. Press)

Carlos I Calle and Karen L Aplin

Chapter 5

The Terrestrial Electrostatic Environment

5.1 The Earth's Atmosphere

The atmosphere of the Earth is a mixture of gases. Close to the surface, the atmosphere is composed of 78% nitrogen, 21% oxygen, and traces of many other gases, including water vapor, argon, and carbon dioxide. This composition changes with altitude. The variations of temperature with altitude are due to the differences in chemical composition in each region of the atmosphere which, in turn, affect the way each region is heated. It is this variation in temperature that divides the atmosphere into distinct regions or layers.

The layer closest to the surface, below about 12 km, is called the *troposphere*. The temperature of the atmosphere in this layer decreases with altitude. The lower end of the troposphere is heated by the Earth's surface. The upper end, away from the surface, is cooler (Figure 5.1). The convection currents that result from this temperature difference are the main cause of terrestrial weather.

Right above it, reaching to an altitude of 40–50 km, lies the *stratosphere*. Part of the oxygen in this layer of the atmosphere is in the form of ozone, O_3, which absorbs ultraviolet (UV) and infrared radiation from the Sun. This direct absorption of the Sun's energy causes a temperature inversion in the stratosphere, with temperature rising with increasing altitude.

The next layer, extending to about 80 km from the surface, is the *mesosphere*. This layer does not contain ozone to absorb the Sun's energy. Atomic oxygen, O, and carbon dioxide radiate energy to space, and the temperature decreases with altitude, reaching about 190 K at the top of the mesosphere.

In the *thermosphere*, which starts at an altitude of about 80 km, temperature again increases with increasing altitude. The extremely low density of this region prevents the formation of oxygen and nitrogen molecules (O_2 and N_2) that are found close to the surface. The atmosphere is heated directly, as atomic oxygen and nitrogen

doi:10.1088/2514-3433/ac878ech5

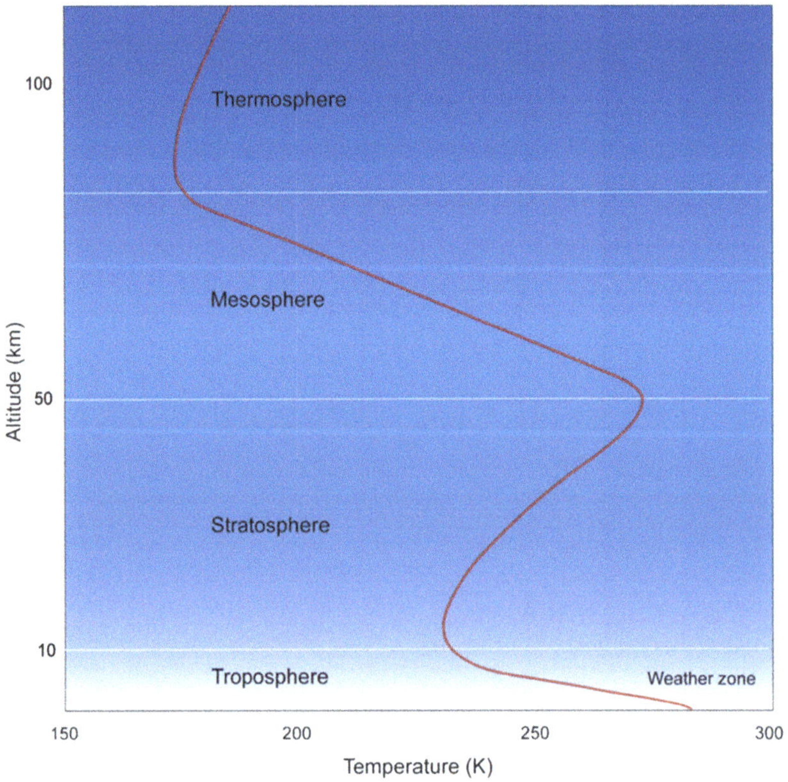

Figure 5.1. Temperature variation (red line) in the Earth's atmosphere.

absorb UV radiation from the Sun. The temperature in the thermosphere increases with altitude, since the most energetic photons from the Sun are absorbed in the upper layers of the atmosphere.

The temperature of the atmosphere at an altitude of 400 km, where the International Space Station (ISS) orbits, is above 750 K. However, the density of the atmosphere is very low in this layer, with too few fast-moving atoms to cause an increase in temperature on the ISS.

The *ionosphere*, which starts in the mesosphere, is the region starting at about 50–70 km and reaching to about 2000 km above the surface where many atmospheric atoms are ionized by solar radiation and cosmic rays. The electrons knocked off from the nitrogen and oxygen atoms in the atmosphere behave as free particles. The highest electron densities exist in the region between 200 km and 400 km. This is the *F* region and it is ionized by the radiation from the Sun during the day and by cosmic rays at night.

The pressure in the atmosphere decreases steadily with altitude throughout these layers. At sea level, the average atmospheric pressure is 1 *atmosphere* (atm), equal to 101,325 N m^{-2} or 1.013×10^5 Pa (= 1.013 bar = 1013 millibar = 14.70 lb in^{-2}). As we go up in the atmosphere, there is less air above us and the pressure decreases. Table 5.1 shows the atmospheric pressure at different altitudes.

Table 5.1. United States Standard Values of Atmospheric Pressure at Different Altitudes.

Altitude	Atmospheric Pressure			
Meters	mbar	kPa	mm Hg	kPa(From barometric formula)
0	1.013	101.33	760	101.33
1000	0.899	89.88	674	90.00
2000	0.795	79.50	596	79.94
3000	0.701	70.12	526	71.01
4000	0.617	61.66	462	63.07
5000	0.541	54.05	405	56.02
6000	0.472	47.22	354	49.76
7000	0.411	41.11	308	44.20
8000	0.357	35.65	267	39.26
9000	0.308	30.80	231	34.87
10000	0.265	26.50	199	30.98
15000	0.121	12.11	90.8	17.13
20000	0.0553	5.529	41.5	9.47
25000	0.0255	2.549	19.1	5.24
30,000	0.0120	1.197	8.98	2.90
40,000	0.00287	0.287	2.15	0.89
50,000	0.00080	0.080	0.60	0.27
60,000	0.00022	0.022	0.16	0.0827
70,000	0.00005	0.005	0.04	0.0253
80,000	0.000011	0.001	0.01	0.0077

5.2 Electrical Breakdown in the Terrestrial Atmosphere

As stated in Chapter 4, electrical breakdown potentials as a function of pressure and electrode gap (Paschen's law) are obtained experimentally. These electrodes are usually parallel-plate electrodes with rounded edges that have a radius of curvature decreasing gradually so that the field has a maximum value at the center of the plane region. Figure 5.2 shows experimental values of the Paschen breakdown potential as a function of pressure multiplied by distance for air.

Figure 5.3 shows experimental values of the breakdown potentials in air near the surface of the Earth for different parallel-plate electrode gaps. With a 1 cm gap, the breakdown potential is about 30 kV, a value often quoted. However, with a 2 cm gap the potential is 57 kV and with a 4 cm gap it is 109 kV, not exactly doubling or quadrupling the value for the 1 cm gap. Figure 5.4 shows how the breakdown potential decreases rapidly as atmospheric pressure decreases with altitude.

5.3 Radiation from the Sun: The Solar Wind

The gas in the Sun's corona has temperatures above 10^6 K, hot enough for particles to escape the Sun's gravity, moving outward at speeds of about 400 km s^{-1}. This

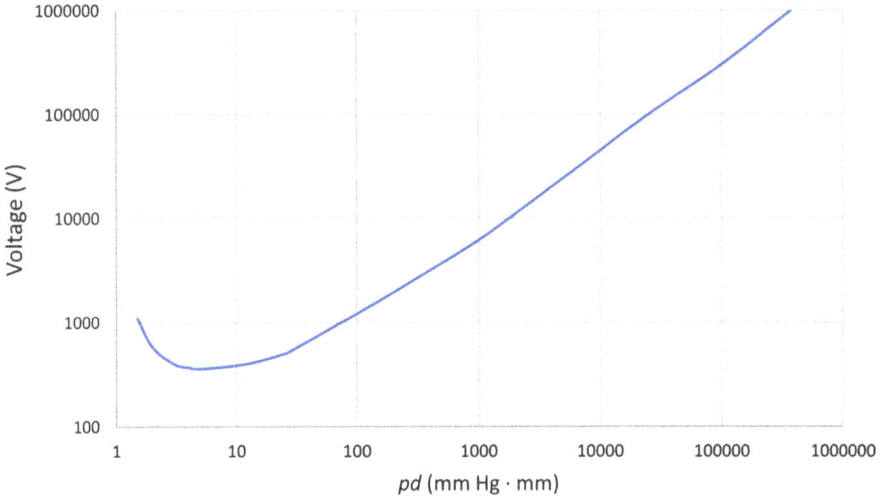

Figure 5.2. Paschen breakdown potentials for parallel-plate electrodes as a function of pressure–distance.

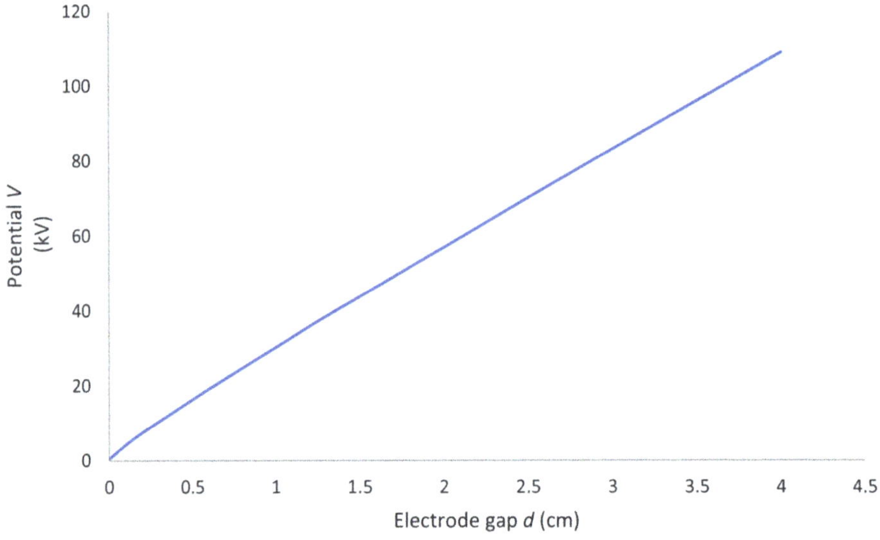

Figure 5.3. Breakdown potentials for several parallel-plate electrode gaps near the surface of the Earth, at 1 atm.

speed makes the solar wind supersonic, since it is faster than the speed of sound in the rarefied gas that exists between the planets. Despite carrying about a million tons of solar matter each second away from the Sun, the particle density in the solar wind in the vicinity of the Earth is very thin, since it drops off as r^{-2}. The energy density of the solar wind at 1 AU is about 5×10^3 eV cm^{-3} (Lewis 1995). The charged solar wind particles are mostly electrons and protons with some helium nuclei and traces (about 0.1%) of heavier ions with energies from about 19 keV to about 100 keV. The solar wind pushes the tails of comets away from the Sun and shapes the magnetic

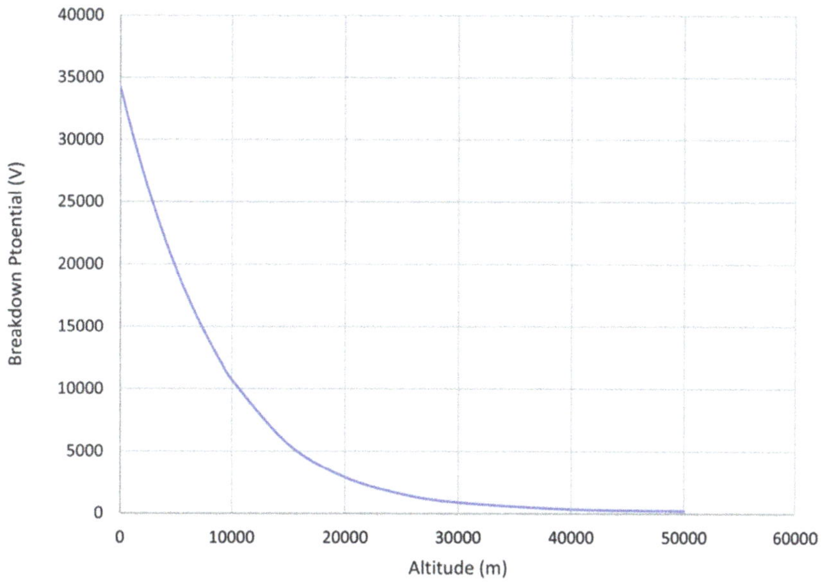

Figure 5.4. Breakdown voltage for different altitudes for a 1 cm gap.

fields around the planets. The solar wind and other emissions of matter from the Sun together form space weather, which was discussed in Chapter 3.

5.4 Radiation Belts

The interaction of the Earth's magnetic field with the solar wind produces a semipermeable barrier to the charged particles in the solar wind called the *magnetosphere*. Most of the solar wind particles flow around the outer edge of the magnetosphere. As they encounter the magnetosphere, the supersonic charged particles in the solar wind form a shock wave (Figure 5.5) that abruptly slows down the supersonic charged particles to subsonic speeds. The front of the bow shock is located at about ten Earth radii, but its exact location depends on changes in solar activity. The magnetosphere stretches out away from the Sun to about 60 Earth radii, forming a long tail that reaches beyond the orbit of the Moon.

When a charged particle from the solar wind enters the Earth's magnetosphere with a velocity **v** at right angles to the magnetic field **B**, the magnetic force acting on the particle will be perpendicular to both the particle's velocity and the magnetic field. This magnetic force is

$$\mathbf{F} = q\mathbf{v} \times \mathbf{B}$$

where q is the particle charge. As the vector product between **v** and **B** indicates, this force accelerates the particle in a direction that is perpendicular to its velocity, making the particle move in a circle at a constant speed. If the charged particle enters the magnetosphere with a velocity component that is parallel to the field **B**,

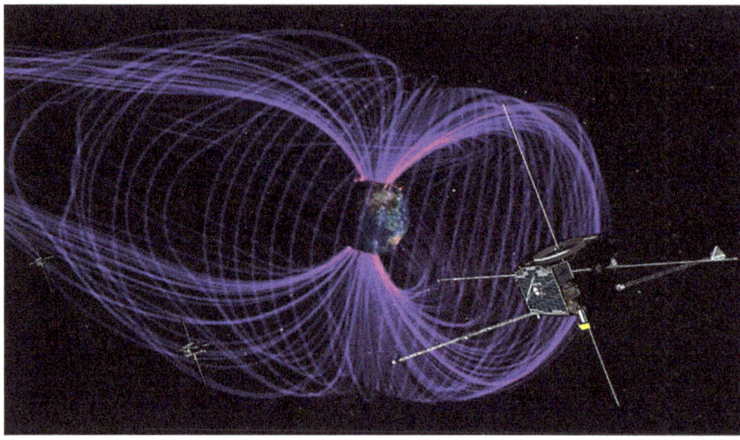

Figure 5.5. The solar wind interacts with the Earth's magnetic field forming a shock wave that slows down the particles. NASA's THEMIS mission has helped map out the magnetosphere with five nearly identical satellites launched since 2007 to orbit the Earth in highly elliptical orbits. (Courtesy of NASA.)

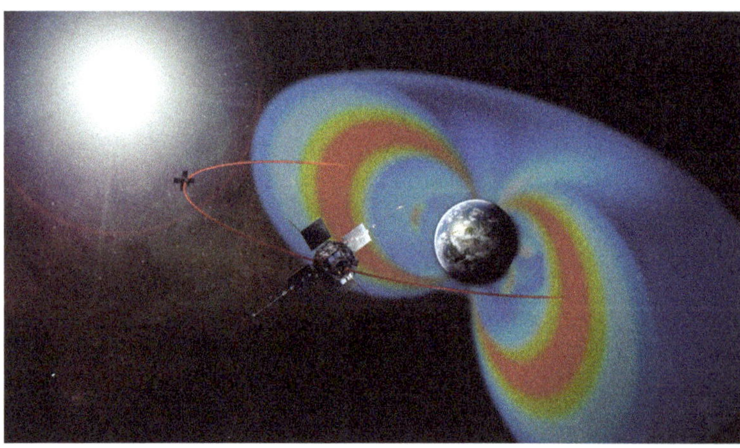

Figure 5.6. The Van Allen radiation belts surrounding the Earth. NASA's Van Allen Probes, launched in 2012, orbit through the belts to study them. (Courtesy of NASA.)

the particle will move along the field but the perpendicular component of its velocity will also force it to move along a circle. The particle will then follow a spiral path.

But, since the Earth's magnetic field is stronger near the poles, the particle will oscillate back and forth in a spiral pattern, trapped in the magnetic field. There are two doughnut-shaped regions of trapped charged particles from the solar wind around the Earth. These are the *Van Allen belts* (Figure 5.6). The inner Van Allen belt, extending from about 2000–5000 km, traps mostly protons. The outer belt, at altitudes between 13,000 km and 19,000 km, traps mostly electrons. Electric and magnetic fields in this belt speed up electrons to nearly the speed of light (Kanekal et al. 2015). The inner edge of the outer belt is a sharp boundary that prevents the most energetic electrons from reaching Earth (Baker et al. 2014).

Figure 5.7. The inner edge of the outer radiation belt (shown in yellow) is a hard boundary that stops the most energetic electrons from entering the gap between the radiation belts. The plasmasphere, shown in green, is sliced open to reveal the inner radiation ring. (Courtesy of NASA.)

Data from NASA's Van Allen Probes mission, launched in August of 2012 to study the radiation belt region (Figure 5.6), has shown the reason for the existence of the hard boundary and of the existence of two radiation belts rather than a single belt. Electrons in the outer belt move at high velocities around the Earth, with only a small velocity component toward the Earth. Other charged particles in the *plasmasphere*—a relatively cool region that starts at about 1000 km and extends into the outer radiation belt—interact with the less energetic electrons at the inner edge of the outer belt causing them to scatter. These scattered electrons repel the drifting high-energy electrons away from the Earth, keeping them out of the gap between the two radiation belts (Figure 5.7).

5.5 Aurora

Violent and sudden eruptions in the solar atmosphere, called *solar flares*, raise its temperature to 5×10^6 K, emitting large quantities of particles and radiation at all wavelengths. *Coronal mass ejections*, lasting several hours, are even more violent, sending more than 10^{12} kg of energetic plasma and radiation into space.

If a solar flare or a coronal mass ejection happens to be directed toward the Earth, the ejected particles interact with the Earth's magnetic field generating magnetic storms that dump some of the charged particles from the radiation belts into the atmosphere. Atoms of nitrogen and oxygen absorb the UV radiation that results from the collisions of these fast charged particles with the atoms in the upper atmosphere. These nitrogen and oxygen atoms then emit this energy as visible light, generating the *aurorae* (*aurora borealis* or *northern lights* and *aurora australis* or *southern lights*). Figure 5.8 shows an aurora photographed from the ISS.

Auroras are a very dynamic phenomenon, undergoing large-scale changes, from being relatively static to having a very active condition several times each night. These abrupt changes are known to be triggered by *substorms*, and cause the auroras

Figure 5.8. Aurora photographed by one of the astronauts of Expedition 6 aboard the ISS on 2003 March 30. (Courtesy of NASA.)

to flicker, sending ions toward the Earth. The existence of these substorms has been known for some time, but their cause has been a mystery. A few years ago, NASA's THEMIS spacecraft fleet, launched in 2007, discovered that particles in the solar wind actually flow along giant *magnetic ropes*, twisted bundles of magnetic field lines that directly connect the Earth's atmosphere to the Sun. More recently, THEMIS researchers proposed that substorms may be caused by bubbles of low-density plasma that ride along these magnetic ropes toward the Earth (Pritchett et al. 2014).

References

Baker, D. N., Jaynes, A. N., Hoxie, V. C., et al. 2014, Natur, 515, 531

Kanekal, S. G., Baker, D. N., Fennell, J., et al. 2015, Conf. Measurement Techniques in Solar and Space Physics (Boulder, CO: American Geophysical Union)

Lewis, J. S. 1995, Physics and Chemistry of the Solar System (San Diego, CA: Academic)

Pritchett, P., Coroniti, F. B., & Nishimura, Y. 2014, JGRA, 119, 4723

Carlos I Calle and Karen L Aplin

Chapter 6

Spacecraft and Satellites in the Electrostatic Environment of the Earth

6.1 Spacecraft and Satellite Orbits

Spacecraft and satellites orbiting the Earth are exposed to the electrostatic environment of near-Earth space. There are three main orbital regions around the Earth: *low-Earth orbit* (LEO), *medium-Earth orbit* (MEO), and *geosynchronous-Earth orbit* (GEO).

LEO is the region with an altitude between 160 km and 2000 km. Most scientific satellites, many weather satellites, and the International Space Station (ISS) are in LEO. The Hubble Space Telescope also orbits in LEO.[1] One special kind of LEO is the *polar orbit*, where satellites orbit the Earth from pole to pole, taking about 99 min to complete an orbit. A satellite in polar orbit will cross from daylight to night-time as it crosses the pole and will view most of the Earth in one day, once at night and once in daylight.

MEOs lie in the space below GEOs (which are at an altitude of 35,786 km) and above 2000 km. These orbits are used for communications, navigation, and science satellites. The Global Navigation Satellite Systems (GNSS) satellites, which include GPS and similar systems, fly on near-circular MEOs at an altitude of 20,200 km, which gives them an orbital period of 12 h. Since a satellite on this orbit will cross over the same spot on the Earth twice in a day, the orbit is called semi-synchronous.

Satellites in GEOs, which have a radius of 42,164 km, orbit at the same rotational speed as that of the Earth. If a GEO is circular with an inclination of zero, the orbit is called *geostationary*. A satellite on a geostationary orbit stays directly above the same spot on the Earth's surface. Geostationary orbits are useful for weather and communications satellites.

[1] The newer James Webb Space Telescope is not in Earth orbit. Instead, it was placed at a special position called the second Lagrange point (L2), named after the French astronomer Joseph-Louis Lagrange. This location places the telescope at a fixed position relative to the Earth and the sun, allowing for a good view of deep space and good line of sight to Earth for data transmission.

doi:10.1088/2514-3433/ac878ech6

6.2 Spacecraft Charging

Spacecraft and satellites in any of these orbits may experience changes in the electrostatic potential at their surfaces relative to the surrounding plasma. Spacecraft charging depends not only on the spacecraft's shape and composition but also on its orbit. Spacecraft surfaces facing the Sun can acquire a net positive electrostatic charge with the release of electrons by photoemission due to UV radiation from the Sun. Since solar UV photons will not interact with the surfaces of the satellite in the shade, differential charging between the sunlit and shadow regions will occur. This differential charging is larger on insulating surfaces. The plasma characteristics in the Earth's near-space environment vary with solar activity and space weather, whereas the variation of UV photon flux over the solar cycle is small and can be neglected.

Charged particles in the plasma from the different regions crossed by a satellite in orbit can also charge the satellite. The amount of charge depends on the energies of the plasma particles in the region, which range from electronvolt to kilo-electronvolt levels. The "wishbone chart" from the *NASA Technical Handbook* shown in Figure 5.1 illustrates the calculated electrostatic surface charging expected for a satellite orbiting at altitudes from 100 to 100,000 km for some latitudes along with the hazard level (NASA 2011). The chart does not include calculations for polar Earth orbits. These calculations were performed for aluminum spherical satellites in the dark (so that photoelectric charging does not need to be included). At latitudes between 40 and 50 degrees north, the potentials reach values of over 400 V, which may produce discharges. The charging asymmetry for the northern and southern latitudes is due to the tilting of the magnetic north pole relative to the geographic pole. The values of the electrostatic potentials on this chart are to be taken as the worst-case surface charging for a spacecraft in the Earth's electrostatic environment (NASA 2011).

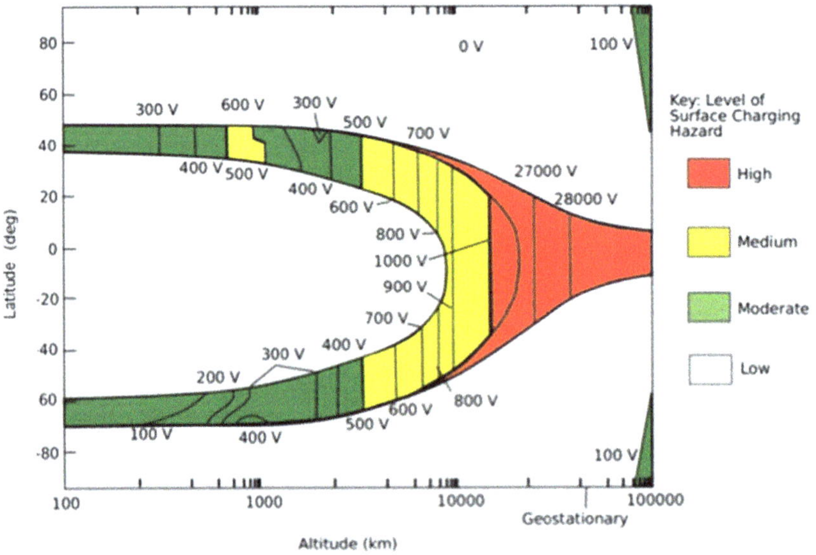

Figure 6.1. Calculated values of the charging potentials for spherical metallic satellites on orbits at some latitudes for altitudes between 100 and 100,000 km. (Courtesy of NASA.)

As Figure 6.1 shows, regions where the electrostatic potentials reach levels that may cause spacecraft charging vary with latitude and altitude. Spacecraft in GEO at certain latitudes are particularly vulnerable, with potentials reaching over 28,000 V.

Spacecraft launched from the surface of the Earth are also exposed to this range of electrostatic environments as they fly to their orbits.

6.3 Spacecraft Charging in LEO

LEO lies in the ionosphere. In this region, at altitudes above 90 km, UV radiation from the Sun splits diatomic oxygen molecules into individual atoms. Solar radiation and cosmic rays ionize the oxygen and nitrogen atoms, producing equal numbers of free electrons and ions. The concentrations of these electrons and ions are relatively high and change with altitude, solar and space weather activity (Ferguson & Hillard 2003). Since electrons have higher mobilities, they are collected more easily, negatively charging spacecraft in this region. Although the LEO plasma density is high, the energy is normally low, at about 0.1 eV. Thus spacecraft charge only to a few volts negative, which is not an issue (Anderson 2012). Figure 6.2 shows

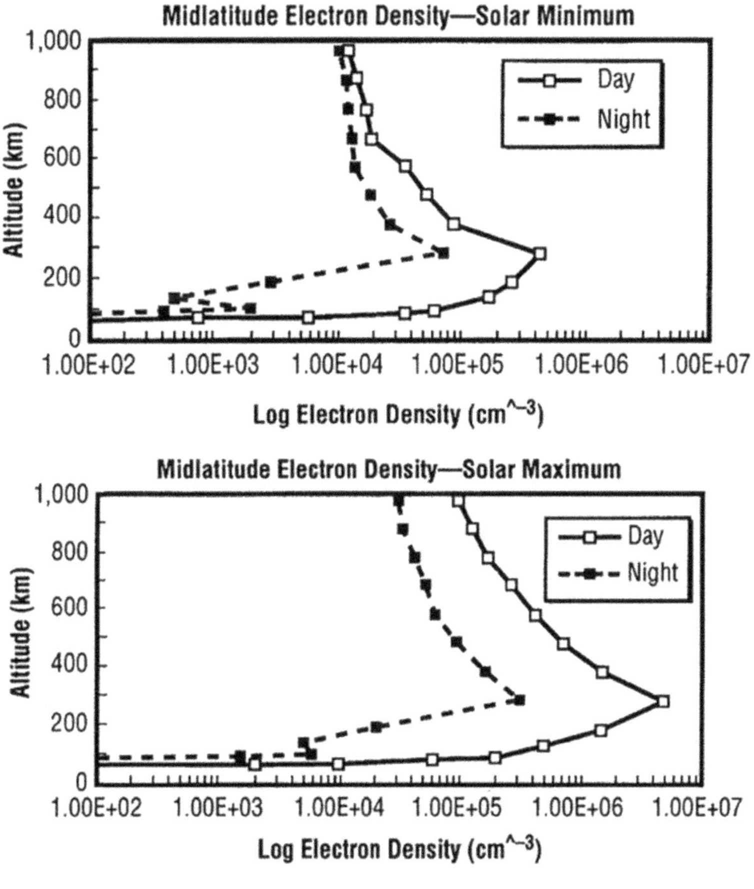

Figure 6.2. Variation in electron density at different altitudes in the ionosphere for a solar minimum (top) and solar maximum (bottom) (James et al. 1994). (Courtesy of NASA.)

the electron density throughout most of the ionosphere for periods with minimum and maximum solar activity (James et al. 1994). The highest electron concentration, at about 300 km, lies in the *F* region of the ionosphere. This region, extending from 200 to 400 km, is where the ISS, the Hubble Space Telescope, and most scientific satellites orbit.

Orbiting spacecraft in the *F* region move at about 8 km s^{-1} in the dense, low-energy plasma. This orbital velocity is slower than the electron thermal velocities, which are about 200 km s^{-1}, but faster than the 1 km s^{-1} thermal velocities of positive ions. Thus, spacecraft in LEO orbit supersonically relative to the ambient ions and subsonically relative to the thermal electrons. As a result, a wake forms behind the spacecraft where the ion density is much smaller than that of the ions in the plasma. This unequal flux results in electrons impacting all LEO spacecraft surfaces while ions can only impact ram surfaces (Ferguson & Hillard 2003; Anderson 2012). However, electron impact is not uniform throughout the spacecraft. The faster moving electrons that enter the wake form a space charge that repels additional electrons, excluding them from the wake region (NASA 2011; Tang et al. 2015).

Spacecraft in polar orbits, even at altitudes of about 300 km, can charge up to several kilovolts due to high-energy electrons reaching those low altitudes. Since the geomagnetic field lines plunge at polar orbits, charging conditions at these altitudes are similar to those in GEO, where spacecraft can charge to potentials in the kilovolt range.

6.4 Charging of the ISS

The ISS orbits the Earth in the *F* region of the ionosphere, at an altitude of 400 km at an inclination of 51.6 degrees. As stated earlier, the relatively high plasma density in LEO limits spacecraft surface charging to a few volts negative. Due to its large size and power requirements, and the inclination of its orbit, the ISS has three additional sources of charge that can be important: the large photovoltaic array system, the magnetic induction process, and the interaction with the high-energy electrons in the polar regions.

Electric power systems on spacecraft normally operate at the 28 V DC standard of the aircraft industry. Interaction with the ionospheric plasma at this relatively low voltage generates low surface charges and is not a concern. However, ISS is not just a spacecraft but a habitat, with much larger power requirements. The ISS photovoltaic power system operates at 160 V to reduce mass and power loss. As is the case for most spacecraft, the solar array system on ISS has the negative end connected to the ISS ground. This set up is due to the scarcity of space qualified power systems with a positive ground when ISS was being constructed. With this arrangement, the positive end of the arrays collects electrons from the dense ionospheric plasma while the negative end collects ions. Although the ion and electron currents are about the same, the electron current density is much higher, and the array and consequently ISS charge negative to values that can reach about −80 V (Koontz et al. 2003). On their own, these array voltages need to be mitigated perhaps by encapsulation of the array edges to prevent arcing due to contamination of the arrays or by bakeout in orbit at 100 °C for about a week to remove contamination (Ferguson & Hillard 2003).

The second source of charge on ISS is the magnetic induction process that results from the motion of the conductive components of ISS through the magnetic field of the Earth. From Faraday's law we know that a magnetic field **B** produces an electromotive force (EMF) that drives a current through a conductor moving with a velocity **v** in that magnetic field. The potential difference of this *motional* EMF is given by

$$\varepsilon_m = \oint \mathbf{v} \times \mathbf{B} \cdot d\mathbf{l}$$

where $d\mathbf{l}$ is an element of length in the conductor pointing in the direction of the current. Long, conducting ISS surfaces moving with a velocity component perpendicular to the Earth's magnetic field can generate these voltages. The effect is larger when ISS orbits through the more concentrated magnetic field lines near the Earth's poles. Potentials of about −90 V relative to the plasma environment have been measured on the photovoltaic arrays (Ferguson & Hillard 2003; Koontz et al. 2003, 2007). These values are not of concern by themselves but, when combined with the negative photovoltaic charging, they can reach values that need to be mitigated.

The third source of ISS charging is due to the daily passage of ISS through the auroral or polar region, where the Earth's magnetic field lines plunge, allowing for higher-energy electrons in the 7–10 keV range to reach LEO altitudes. Measurements of the satellites of the Defense Meteorological Satellite Program, which orbit in the auroral region at an altitude of 840 km, show that they can charge to −2 kV when they encounter intense electron flux (Anderson 2012). At the lower altitude at which ISS orbits, the plasma density is much higher and high-level charging should not normally take place. However, the plasma density in the wake can be up to two orders of magnitude lower (Anderson 2012). High-energy electrons in the auroral regions can charge surfaces in this area. Data from the US Department of Defense/European Space Agency/Russian Roscosmos State Corporation satellite, suggest that the probability for this charging effect is low but not zero (Koontz et al. 2007).

To mitigate these problems, NASA installed a system to ground ISS to the ambient plasma. This system, called the *plasma contactor unit* (PCU), generates a high-density plasma that makes electrical contact with the ambient plasma (Carpenter 2004). The system uses a hollow cathode that converts a small supply of xenon gas into ions and electrons that are discharged to space, thus carrying away the excess charge (Figure 6.3). With the PCU in place, voltages on ISS conducting surfaces do not exceed ±20 V (Ferguson & Hillard 2003). For redundancy, two PCUs were installed on the Z1 truss of ISS. The units operate only during astronaut extra-vehicular activities and during spacecraft docking.

Insulator surfaces on the ISS in general do not build up charges due to the neutralizing effect of the surrounding plasma. This neutralizing effect does not happen in the wake, where, as stated above, the plasma density can be two orders of

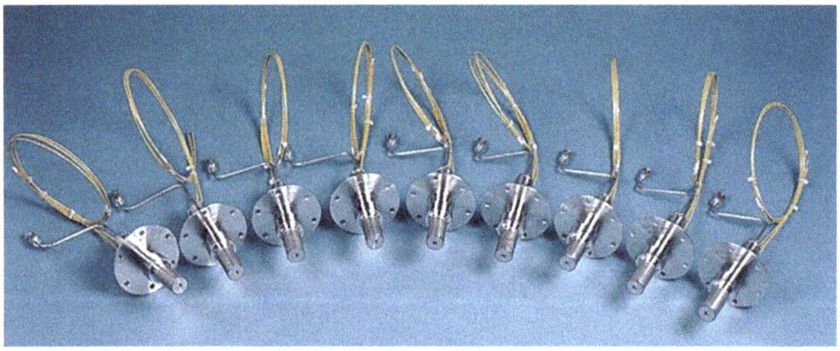

Figure 6.3. The hollow cathode assembly in the PCU on the ISS. (Courtesy of NASA.)

magnitude lower. In the unprotected region of the wake, differential charging can develop on insulating surfaces when the ISS passes through the auroral region, where it is exposed to high-energy electrons.

On 2009 May 11, NASA launched the Space Shuttle to repair and replace several instruments aboard the Hubble Space Telescope, which orbits in LEO. One of the repair tasks was to replace the low-voltage power supply board for the Space Telescope Imaging Spectrograph. This task required the astronauts to expose the replacement board to the space environment and to possible triboelectric charge build-up on astronauts' suits and gloves (due to possible contact and separation of suit and glove surfaces). All four Hubble Servicing Missions required the astronauts to use the Shuttle robotic arm to position the telescope in the Shuttle's payload bay. Thus this repair was going to be performed in the wake, where the plasma density is not high enough to neutralize triboelectrically generated charges on those insulating surfaces. Extensive high-vacuum testing at NASA showed that triboelectric charging at levels that could generate discharges damaging to the boards were possible (Buhler et al. 2012). As a result, an extended grabber tool was developed to handle the boards. There were no discharge incidents during the mission.

6.5 Spacecraft Charging in MEO

Although most satellites are placed in LEO or GEO, there are currently about 70 satellites orbiting in MEO (Horne 2015). Since the Van Allen radiation belts occupy the region between 2000 km and 19,000 km, most MEO satellites are GNSS satellites orbiting just beyond the belts, about 20,000 km up. However, several scientific satellites, such as the NASA THEMIS mission satellites, have highly elliptical orbits through the MEO region and beyond (Figure 6.4). High-speed electrons in the outer radiation belt are a hazard for these satellites as well as for any spacecraft crossing the radiation belts on the way to Mars or to other solar system destinations.

As the solar wind stretches the long tail of the magnetosphere past the orbit of the Moon, it becomes overly stretched, snapping back, accelerating plasma electrons with it. These fast-moving electrons emit electromagnetic waves called *chorus* waves, because they sound like singing birds when the signal is played through

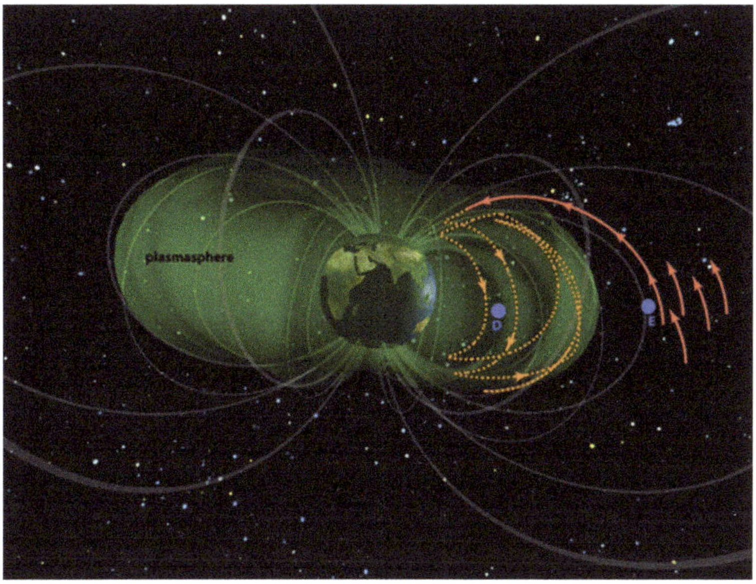

Figure 6.4. Elliptical orbits of THEMIS *D* and *E*, two of the five NASA satellites mapping the magnetosphere. These satellites discovered that chorus waves (red lines) generated by the magnetotail change into hiss waves (yellow lines) that kill energetic electrons from the outer radiation belt. (Courtesy of NASA.)

loudspeakers. These chorus waves accelerate electrons in the outer radiation belt to near the speed of light. Two of the THEMIS satellites (*D* and *E*) discovered that when the chorus waves interact with the plasma in the radiation belt, they lose energy and turn into *hiss* waves, so named because they make that sound when played through audio equipment. Hiss waves deflect high-energy electrons in the upper radiation belt toward the Earth, where they lose energy as they interact with atoms in the atmosphere (Li et al. 2010).

In MEO, an issue that is probably more important than spacecraft charging is *deep dielectric charging*. Electrons and ions with sufficient energies can penetrate spacecraft surfaces, causing internal disruptions in electronics systems. Electrons with energies of a few kilo-electronvolts can penetrate about 1 μm into a dielectric (Tascione 1988). Electrons with energies above 50 keV can penetrate spacecraft metallic surfaces (Robinson & Coakley 1992). In the radiation belts, electron energy can reach the mega-electronvolt range. Figure 6.5 indicates the mean penetration energies of protons and electrons into aluminum (NASA 2011).

6.6 Spacecraft Charging in GEO

GEO is the most important region for spacecraft charging because the plasma energy can be high and because of the high density of satellites in this region (about 40% of all satellites are in GEO). GEO also resides in the magnetosphere but outside the radiation belts. Satellites in GEO are close to the *magnetopause*, the boundary between the Earth's magnetic field and the solar wind. During space weather

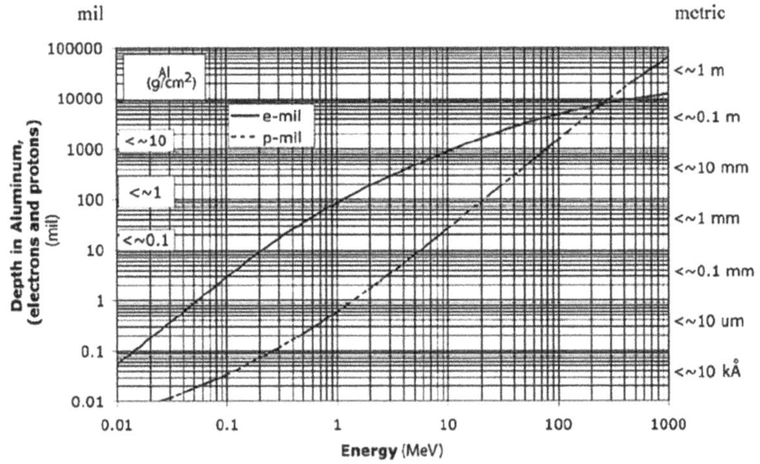

Figure 6.5. Penetration depth into an aluminum slab as a function of energy for electrons (solid line) and protons (dotted line) (NASA 2011). (Courtesy of NASA.)

activity, the front of the magnetosphere is pushed in past the geosynchronous orbits, a phenomenon that takes place several times a year.

On the night side of the Earth, at opposite ends to the magnetosphere's nose, spacecraft are immersed in the night side plasma. Due to its low density (about 1 elementary charge per cm^3), this high-energy plasma is not able to neutralize any charge that builds up on spacecraft surfaces. Spacecraft surfaces in darkness can reach potentials between −10 kV and −30 kV. On the lit side, photoelectron emission due to the incident solar UV radiation results in surface potentials of +2 V to +5 V.

The large potential difference between the lit and dark sides of a spacecraft can cause differential charging. If the skin of the spacecraft is not electrically conductive, a potential difference of about 2 kV to about 11 kV builds up between the two sides. For aluminum or other electrically conductive surfaces, the potential difference is lower, of the order of 1 kV or less.

Of greater concern in this region is deep dielectric charging. Electrons with energies above 10 keV can penetrate insulating materials or pass through insulating films. A large charge build-up in an insulator can result in a rapid discharge that may damage onboard equipment.

6.7 Mitigation Techniques

Space agencies and spacecraft designers around the world have developed several techniques to mitigate spacecraft charging. Spacecraft can have metallic surfaces, usually made of aluminum, as well as electrically insulating surfaces. The NASA Technical Handbook "Mitigating in-space charging effects" (NASA 2011) has charge mitigation recommendations for both metallic and non-metallic spacecraft surfaces. For metallic surfaces, NASA (2011) recommends bonding (grounding) all metallic structural elements to the chassis with a leakage resistance of less than 108 ohms. In addition, all structural and mechanical parts should be electrically bonded to each other with a DC current resistance of less than 2.5×10^{-3} ohms at each joint. For non-conducting or dielectric surfaces, NASA (2011) recommends conducting extensive analyses of the dissipation of charge to ground for all non-conductors to be used, to show that they will not present a hazard to the spacecraft. The GEO Spacecraft Charging Guidelines (Purvis et al. 1984) recommends coating all non-conductive surface materials on a spacecraft with a conductive material that can be grounded.

Dielectric materials used inside the spacecraft, such as wire insulation and circuit boards, should be made statically dissipative; that is, they should be able to leak charge faster than they can acquire it. These statically dissipative materials should be bonded to the spacecraft structure. Indium tin oxide (ITO) is a common statically dissipative coating that can be used on many dielectrics. ITO has the advantage of being transparent (it is commonly used in touch screens).

For large spacecraft—such as the ISS—plasma contactors are an effective method to control charging, as mentioned earlier. Simpler systems to balance the potential of spacecraft by releasing electrons into a vacuum, removing the need for a gas, have also been developed. One form of emitter uses *thermionic emission*, electrons emitted from a heated electrode (Bekkeng et al. 2019). More novel devices use micrometer-sized tips and holes connected to a bias potential, exploiting a quantum tunneling phenomenon called *field emission* (where electrons are released from a material subjected to a high electric field). Prototype devices have been tested in the lab, but are not yet in place (Frigot et al. 2007; Aplin et al. 2009).

References

Anderson, P. C. 2012, JGR, 117, A07308

Aplin, K. L., Kent, B. J., Song, W., & Castelli, C. 2009, AcAau, 64, 857

Bekkeng, T. A., Helgeby, E. S., Pedersen, A., et al. 2019, ITAES, 55, 2951

Buhler, C. R., Clements, J. S., & Calle, C. I. 2012, Proc. 2012 Joint Electrostatics Conf. (Cambridge, ON: Electrostatics Society of America)

Carpenter, C. B. 2004, On the operational status of the ISS plasma contactor hollow cathodes NASA Technical Report NASA/CR-2004-213184

Ferguson, D. C., & Hillard, G. B. 2003, Low Earth orbit spacecraft charging design and guidelines, NASA Technical Report, 2003-212287

Frigot, P.-E., Aplin, K. L., Kent, B. J., Nicolini, D., & del Amo, J. A. G. 2007, Proc. 30th Int. Electric Propulsion Conf. (Florence: Electric Rocket Propulsion Society)

Horne, R. B. 2015, SEREN Workshop on Physical Pathways to Space Weather Impacts (London: Space Weather Research Network)

James, B. F., Norton, O. W., & Alexander, M. B. 1994, The natural space environment: effects on spacecraft, NASA Reference Publication 1350

Koontz, S., Valentine, M., Keeping, T., et al. 2003, 8th Spacecraft Charging Technology Conf. (NASA Marshall Space Flight Center)

Koontz, S., Suggs, R., Schneider, T., et al. 2007, International Space Development Conf. (NASA)

Li, W., Thorne, R. M., Nishimura, Y., et al. 2010, JGR, 115, A00F11

NASA, 2011, Mitigating in-space charging effects—a guideline, NASA Technical Handbook NASA-HDBK-4002A
Purvis, C. K., Garrett, H. B., Whittlesey, A. C., & Stevens, N. J. 1984, NASA Technical Paper, 236
Robinson, P. A., & Coakley, P. 1992, ITEI, 27, 944
Tang, D., Yang, S., Zheng, K., et al. 2015, PlST, 17, 288
Tascione, T. E. 1988, Introduction to the Space Environment (Malabar, FL: Orbit)

AAS | IOP Astronomy

Electrostatic Phenomena on Planetary Surfaces
(Second Edition)

Carlos I Calle and Karen L Aplin

Chapter 7

The Electrostatic Environment of the Moon

The moon has an extremely rarefied atmosphere, composed of small amounts of helium, argon, neon, ammonia, methane, and carbon dioxide at a pressure of 10^{-12} mbar. This value represents a good vacuum. It is about the same as the vacuum found where the International Space Station (ISS) orbits. With a molecular gas concentration of 10^6 cm^{-3} near the surface (10^{13} times smaller than the number of molecules near the surface of the Earth), the lunar atmosphere has no influence on the electrostatic environment of the Moon. What partly determines this environment is the flux of charged particles from the solar wind and cosmic rays. The interaction of this flux with the regolith produces a complex electrostatic environment.

7.1 The Lunar Surface Environment

The lunar surface is covered by a regolith composed of several meters of rock and granular material ranging in size from a few centimeters to a few nanometers in diameter. Meteoroid impacts have ground the top layers of the regolith down to small particles. The average size of the Apollo dust samples is about 70 μm. The regolith also includes a significant fraction of dust in the micrometer and sub-micrometer scale (Grün et al. 2011). However, the size distribution is not homogeneous over the surface of the Moon, as evidenced by Apollo images (Figure 7.1). The variation in size distribution is due to the different lengths of exposure to the dynamic process of meteorite impact. The Apollo 15 landing site, for example, was on the eastern edge of Mare Imbrium, in the fine dust of the impact ejecta. The Apollo 16 landing site was in the lunar highlands, where rocks rich in the mineral feldspar are found.

7.2 The Lunar Electrostatic Environment

Since the Moon does not have a substantial atmosphere, the lunar surface is directly exposed to the solar wind, to cosmic rays, and to solar UV radiation. The solar wind

doi:10.1088/2514-3433/ac878ech7

Figure 7.1. Differences in regolith size distribution are evident in these images of the Apollo 15 and Apollo 16 landing sites. (Courtesy of NASA.)

plasma is a flow of electrons and ions with a density of about 5 elementary charges cm^{-3} moving at about 400 km s^{-1} (Dessler 1967). The density of the cosmic rays is about six orders of magnitude smaller. The formalism describing the complex lunar electrostatic environment that results from these fluxes was partially developed by Manka (1973) and completed by Halekas et al. (2005) and Farrell et al. (2007). The basic premise of this formalism is simply the law of conservation of charge: in equilibrium, the lunar surface will develop a charge so that the sum of all current fluxes cancels out. These current fluxes are: electrons and protons in the solar wind; protons (~90%), alpha particles (~9%), electrons (~1%), and a small fraction of heavy nuclei in cosmic rays; photoelectrons (Feuerbacher et al. 1972); reflected and back-scattered electrons; secondary electrons from the lunar surface; solar wind ions reflected from the lunar surface (McComas et al. 2009; Saito et al. 2008; Weiser et al. 2010); and galactic cosmic rays (GCRs) partially reflected off the lunar surface (Porter et al. 2013). Electrostatic potential contributions from these currents have been calculated using the Manka formalism for typical solar wind conditions using data from the Electron Reflectometer (ER) instrument on NASA's Lunar Prospector mission (Halekas et al. 2005). The ER instrument provided data on electron concentrations and temperatures. The plasma environment around Lunar Prospector's orbit was assumed to be neutral and the electrons and ions were assumed to have the same temperatures. These considerations showed that the majority of the contribution to lunar charging comes from only four currents: photoelectrons, solar wind electrons, solar wind ions, and secondary electrons (Stubbs et al. 2005).

The amount and sign of the charge that develops on the lunar surface is different for the sunlit and the night side of the Moon. On the sunlit side, the photoelectrons emitted by the sub-200 nm UV radiation from the Sun as well as by solar X-rays leave the day side surface of the Moon positively charged. Some photoelectrons escape and join the plasma electrons surrounding the Moon. However, most photoelectrons have insufficient energy to escape, forming a sheath with a density of 10^3–10^4 electrons cm^{-3} about 1 m above the daylight lunar surface that shields the surface from the plasma (Singer & Walker 1962). A small positive potential of about 5–10 V would balance the photoelectron and incident solar wind electron currents

(Halekas et al. 2012). Because of the shielding effect of the photoelectron sheath, the photoelectron current dominates on the sunlit side for the Moon.

On the night side, in the solar wind wake, the plasma density is substantially reduced, resulting in an increase in electron kinetic energy. Electrons in the solar wind plasma, moving faster than the ions, become the dominant current. This electron flux charges the lunar surface to potentials of about −50 to −200 V (similar to electron temperature) (Halekas et al. 2002). When the Moon crosses the Earth's plasma the potentials on the night side reach values of several hundred kilovolts. The Lunar Prospector spacecraft measured night side surface potentials of −4.5 kV during periods of intense solar activity (Halekas et al. 2009). A sheath extending to about 1 km from the surface is then formed on the lunar night side (Freeman & Ibrahim 1975; Singer & Walker 1962).

A similar but smaller plasma wake is formed on the dark side of a crater in the lunar polar region where the solar wind flows horizontally over the crater. As on the night side, energetic plasma electrons move into this wake ahead of the ions, charging the surface negatively to hundreds of volts (Farrell et al. 2008).

The lunar surface is then enveloped in a Debye sheath with a Debye length of the order of meters on the day side and of kilometers on the night side that shields the lunar surface potential (Figure 7.2). Recall from Chapter 2 that the Debye length is the distance beyond which the electrostatic field of a charge is shielded by the opposite charges it has attracted.

The electrostatic fields near the *lunar terminator*, the boundary between the day and night sides, are very complex. In this region, the surface charge changes from positive in the day side to negative in the night side. Lunar Prospector measurements showed that the transition from the photoelectron sheath to the solar-wind plasma

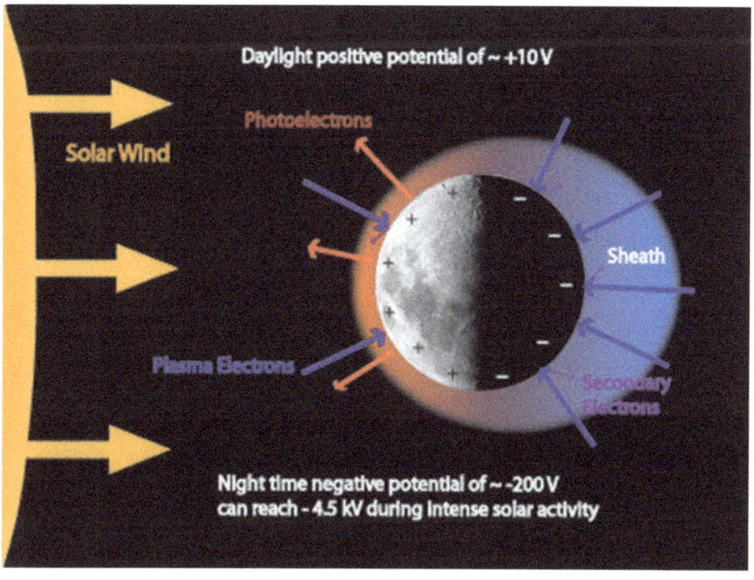

Figure 7.2. Lunar electrostatic environment.

sheath takes place at a subsolar angle of about 67°. At this location, the outward photoelectron emission current balances the inward solar wind electron current and the potential changes from positive to negative. At the terminator (90° subsolar angle), the potential is about −40 V (Farrell et al. 2007; Stubbs et al. 2005).

Although fairly small compared to the photoelectron and plasma electron currents, secondary emissions should still be considered. Secondary radiation is due to interactions between incident GCRs and the atoms in the lunar regolith. The Cosmic Ray Telescope for the Effects of Radiation (CRaTER), an instrument on NASA's Lunar Reconnaissance Orbiter (LRO), measured space radiation at the Moon to determine protection levels for human missions. These measurements were performed during the deep solar minimum at the beginning of the mission in 2009. Many of the energetic electrons and nuclei in GCRs are deflected by the Sun's magnetic field. However, during a solar minimum, the weakened solar wind does not carry the solar magnetic field as far, allowing for a larger flux of GCRs which, in turn, generate a flux of secondary radiation about 30%–40% higher than expected (Schwadron et al. 2015). This amount is relatively small and is comparable to the annual flux experienced by X-ray technicians.

7.3 Electrostatic Charging of the Lunar Regolith

GCR particle radiation as well as solar energetic particle events (SEP) incident on the lunar surface can generate deep dielectric charging of the regolith down to a depth of about 1 m (Schwadron et al. 2015). Scientists expect this process to be of particular importance in the lunar polar regions. Since the lunar regolith has a very low electrical conductivity, electrostatic discharges can take place through the affected section of the regolith. NASA's Acceleration, Reconnection, Turbulence and Electrodynamics of the Moon's Interaction with the Sun (ARTEMIS) mission[1] (Figure 7.3), launched in 2010, is measuring the solar radiation incident on the lunar surface as the Moon moves in and out of the Earth's magnetic field. Reka Winslow and collaborators estimated the potential difference between ARTEMIS and the lunar surface during two strong solar energetic particle events. They then used this value with a deep dielectric charging model developed by Jordan et al. (2015a) to estimate the subsurface electric field strength and the possibility that dielectric breakdown had occurred in the lunar regolith during those events (Winslow et al. 2015). These results can be combined with the occurrence of solar energetic particle events to predict the frequency of dielectric breakdown throughout different regions of the Moon and in particular, in the polar regions (Jordan et al. 2015b). GCRs can also alter the chemical composition of the regolith (Jordan et al. 2015a).

The complex lunar electrostatic environment affects surface dust behavior in unexpected ways. There has been some evidence of a horizon glow on the Moon that has been interpreted to be due to dust levitation and dust transport. This phenomenon was first observed during the NASA Surveyor missions to the Moon in the 1960s. The

[1] This ARTEMIS mission is different from the newer NASA Artemis mission that aims to establish a sustainable presence on the moon to prepare for missions to Mars.

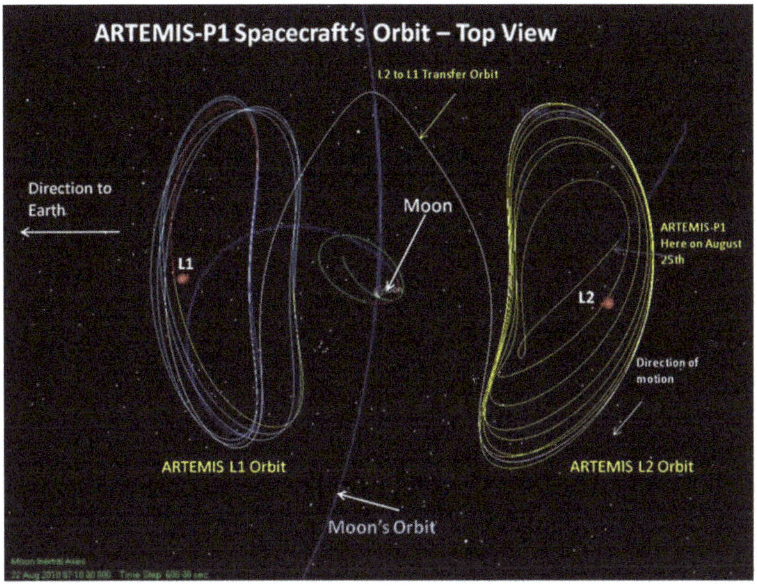

Figure 7.3. The ARTEMIS spacecraft's orbit. (Courtesy of NASA.)

Surveyor 5, 6, and 7 lunar landers obtained television images of a glow just above the lunar horizon (Criswell 1973). The optical properties of this lunar horizon glow implied a relatively high particle concentration that could not be attributed to meteorite impacts. It was proposed instead that the glow was due to scattered sunlight from a cloud of submicron dust particles levitated by electrostatic forces up to a height of about 1 m above the surface near the terminator (Rennilson & Criswell 1974).

Additional evidence of this horizon glow and of dust levitation came during the Apollo missions. Apollo 17 astronauts sketched observations from orbit of what they referred to as "streamers", "bands", or "twilight rays" on the lunar horizon (Figure 7.4; Criswell 1973). A more recent analysis of these sketches puts the radius of the particles at less than 10 μm (Stubbs et al. 2006). The Lunar Ejecta and Meteorites (LEAM) experiment on Apollo 17 detected the presence of dust clouds as well as evidence of slow-moving, highly charged dust particles (Berg et al. 1975).

The 1994 Clementine mission contained a dedicated Lunar Horizon Glow experiment that required the spacecraft to be in the lunar umbra. Unfortunately, due to the geometry of the trajectory relative to the Sun, conflicts with the more important mapping experiments onboard and data downlinks prevented the Lunar Horizon Glow experiment from being carried out (Sorensen & Spudis 2005). However, images with the spacecraft's tracker camera detected a faint glow above the brighter glow of zodiacal dust particles (Figure 7.5). The interpretation of these images is complicated and has never been completed satisfactorily (Hahn et al. 2002).

NASA's Lunar Atmosphere and Dust Environment Explorer (LADEE), which launched in September 2014 to study the Moon from orbit until it was intentionally crashed on the lunar surface six months later, discovered that the Moon is surrounded by a permanent dust cloud lofted from the surface by the impacts of interplanetary dust

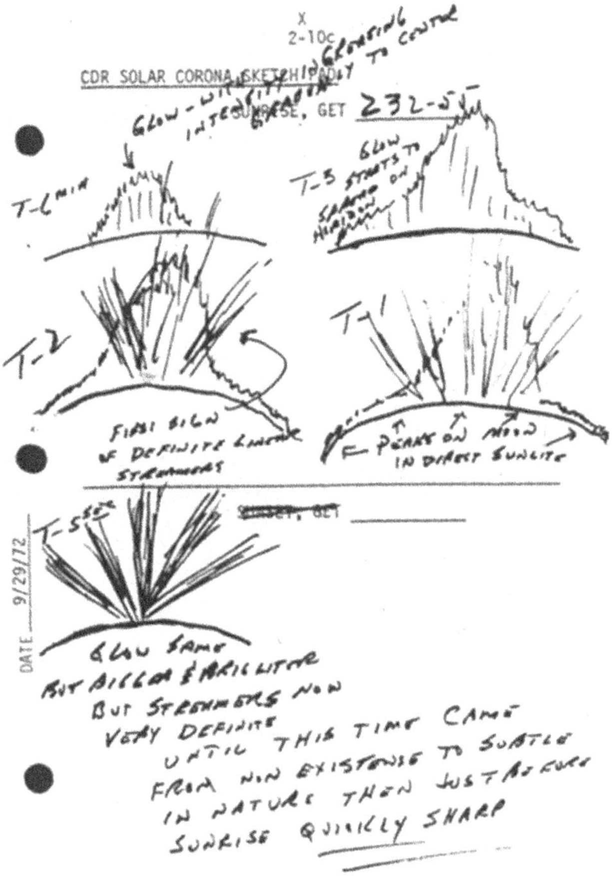

Figure 7.4. Sketches drawn in lunar orbit by Apollo 17 astronauts showing streamers reaching up to about 100 km from the surface. (Courtesy of NASA.)

particles. These particles strike the surface at speeds of 34 km s^{-1}, vaporizing part of the soil and releasing a large amount of heat. A single interplanetary dust particle can kick up thousands of surface dust particles. The total estimated mass of the lunar dust cloud is 120 kg. However, LADEE did not find evidence for the horizon glow discovered by Surveyor spacecraft and the Apollo 17 mission. Submicron particle densities were measured at less than 100 particles m^{-3}, a level considered to be too low to produce a horizon glow (Szalay & Horanyi 2015). Thus the issue of the electrostatic lofting of submicron dust particles generating a horizon glow on the Moon remains unresolved.

7.4 Triboelectric Charging on the Lunar Surface

An object moving on the surface of the Moon will charge due to contact with the surface as well as its exposure to the surrounding plasma environment. Given the possible high potentials on the lunar night side during periods of intense solar activity, this triboelectric charging may affect human missions to the lunar surface.

Figure 7.5. Faint glow above the brighter glow caused by interplanetary dust particles observed during the Clementine mission. A definite interpretation of this faint glow has not been achieved. (Courtesy of NASA.)

A recent study by Jackson et al. (2015) indicates that the time it takes for charge to dissipate depends on the local plasma environment. They performed modeling analyses with the rubber tires on the Modular Equipment Transporter (MET) used during the Apollo 14 mission and with the aluminum and titanium tires on the Lunar Roving Vehicle (LRV) used during Apollo 15, 16, and 17.

On the lunar day side, a rover is exposed to the same electrostatic environment as the lunar surface, emitting photoelectrons with energies that depend on the work function of the material. Calculations for the MET insulating wheel under lunar daylight conditions showed that it would charge to a potential similar to the ~3 V potential of the lunar surface. The metallic wheel on the LRV would charge to 4.62 V relative to the local surface potential. These low potentials would decay very quickly (in less than 0.5 ms) to reach equilibrium with the surrounding environment (Jackson et al. 2015). As the rover moves along the lunar surface, repeated wheel–regolith contact and separation would cause triboelectric charging on both surfaces. Results from modeling the motion of the two wheels using a charge continuity equation developed by Farrell et al. (2010) for several speeds ranging from 0.01 m s^{-1} to 3 m s^{-1} show that in the photoelectron dominated daylight plasma region, the charge dissipates almost instantly in all cases. The photoelectron plasma region is conductive and triboelectric charging on the lunar day side is not an issue.

On the lunar night side, where surface potentials can regularly reach −200 V, a rover will be exposed to the more energetic solar wind electrons that dominate this region as well as to secondary electron currents. A similar situation occurs for a rover moving on the shadowed polar crater region, where the higher energy solar wind electrons and secondary electrons from the surface charge the surface negatively. At the higher electron speeds in these two regions, both types of wheels would charge to about −1 kV, dissipating to equilibrium in about 10^7–10^8 s (Jackson et al. 2015). These large potentials and long dissipation times will pose a hazard to astronauts and equipment during lunar missions.

7.5 Planned Measurements with Lunar Landers

As was mentioned in Section 7.3, the density of the photoelectron sheet used in the lunar electrostatic models used measurements taken from orbit with the Electron Reflectivity instrument on Lunar Prospector. NASA and the University of Colorado are developing a low frequency radio spectrometer that will deploy antennas located to measure the photoelectron sheet density from 1 to 3 m above the lunar surface. The instrument, called Radio wave Observations at the Lunar Surface photoElectron Sheath (ROLSES), will have a frequency range from 10 kHz to 30 MHz.

A model developed by Poppe and Horanyi (2010) predicted the plasma density at a height of 1 m to be 6×10^7 m^{-3} which remains constant to 3 m above the surface. A second model by Farrell et al. (2013) predicted values of 5×10^7 m^{-3} at 1m and 1×10^7 m^{-3} at 3 m, respectively. ROLSES will provide direct measurements of those values. This instrument should also be able to detect radio emissions from the Sun, from Jupiter, and from Earth's auroral bursts. ROLSES is scheduled to launch in December 2022 on the Nova-C lander provided by Intuitive Machines, a company participating in NASA's Commercial Lunar Payload Services (CLPS) program.

NASA, the University of Central Florida, and Lunar Outpost, a company that is also on the CLPS program, are developing electrostatic sensors that will be installed on the wheel of a rover to measure the electrostatic charging of several wheel components as well as that of other space materials as the wheel rolls on the lunar soil. Measurements will be taken during the lunar day, lunar night, and on crater shadowed regions. These measurements should serve to validate the laboratory experiments performed by Jackson that were described in the previous section (Jackson et al. 2015).

7.6 Dust Removal from Surfaces with Dielectrophoretic Forces

Lunar dust particles will acquire a charge via the triboelectric mechanism when in contact with rovers or astronauts during exploration activities. In the energetic electron plasma of the lunar night, this triboelectric charge will take a long time to dissipate.

However, in the conductive plasma of the lunar day, this charge will quickly dissipate. Electrostatically charged lunar dust will tenaciously adhere to spacesuits, astronaut boots, camera lenses, astronaut helmet visors, seals on habitat hatches, and other equipment. Dust may remain on the surfaces even when its charge dissipates, due to Van der Waals forces between small dust particles and smooth surfaces. Neutral dust particles may also get trapped in the spacesuit fabric. NASA is working on a technology that uses electrostatic and dielectrophoretic forces to remove dust from those devices.

The Electrodynamic Dust Shield (EDS) technology is based on a concept developed at NASA in 1967 (Tatom et al. 1967), advanced further at the University of Tokyo for clean room applications (Aoyoma & Masuda 1971; Masuda 1970; Masuda et al. 1972, 1988; Masuda & Matsumoto 1973), and implemented for dusty planetary environments by NASA Kennedy Space Center (Calle et al. 2007, 2011, 2009). This technology removes dust from surfaces with forces generated by electric fields from thin electrodes embedded under the surface (Figure 7.6). These external electric fields will cause positive and negative charges

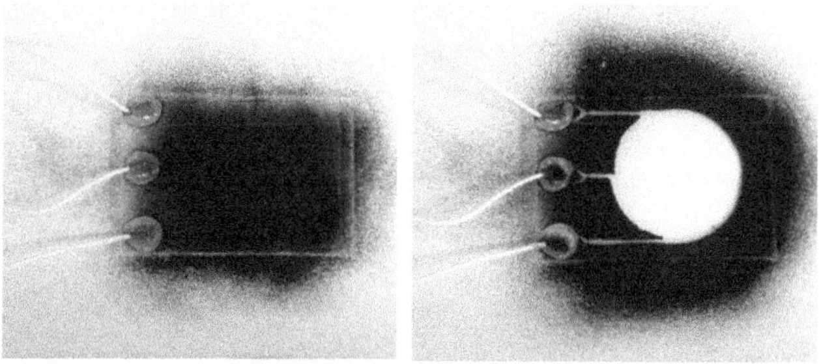

Figure 7.6. (Left) Lunar simulant dust sprinkled over an EDS with transparent indium tin oxide electrodes on a glass panel. (Right) When activated, dielectrophoretic forces remove dust from the area containing the transparent electrodes. (Courtesy of NASA.)

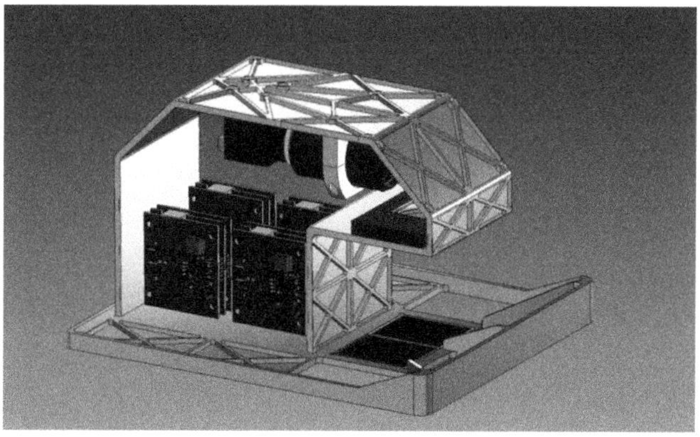

Figure 7.7. Concept drawing of the EDS lunar demonstration payload. A glass EDS in a polymer frame (shown in brown) protects a camera enclosure. The two EDS panels at the base of the payload will be imaged by the camera. All three EDS panels will be activated by the electronics boards on the left side of the enclosure. (Courtesy of NASA.)

usually present even on neutral particles to migrate to opposite ends, polarizing the particle. The EDS electric field is non-uniform with field lines that point away from an electrode with positive potential, curving toward the adjacent electrode at a negative potential. This non-uniform field generates a dielectrophoretic force that moves the polarized particles.

After extended exposure of several EDS panels to the space environment on an external facility on the ISS, NASA is preparing a payload to demonstrate the technology on the lunar surface. The payload will fly to the Moon on a Firefly Aerospace lander as part of NASA's CLPS program (Figure 7.7).

References

Aoyoma, M., & Masuda, S. 1971, Proc. General Conf. of the Institute of Electronic Engineers in Japan, Vol. 821

Berg, O. E., Wolf, H., & Rhee, J. 1975, Interplanetary and Zodiacal Light (Heidelberg: Springer), 233

Calle, C., Buhler, C., Johansen, M., Hogue, M., & Snyder, S. 2011, AcAau, 69, 1082

Calle, C., Buhler, C., McFall, J., & Snyder, S. 2009, Journal of Electrostatics, 67, 89

Calle, C. I., Mazumder, M. K., Immer, C. D., et al. 2007, Controlled particle removal from surfaces by electrodynamic methods for terrestrial, lunar, and Martian environmental conditions, Proc. Electrostatics 2007

Criswell, D. R. 1973, Photon and Particle Interactions in Space, ed. R. J. L. Gerard (Dordrecht: Reidel), 545

Dessler, A. J. 1967, RvGeo, 5, 1

Farrell, W. M., Stubbs, T. J., Delory, G. T., et al. 2008, GeoRL, 35, L19104

Farrell, W. M., Stubbs, T. J., Halekas, J. S., et al. 2010, JGR, 115, E03004

Farrell, W. M., Stubbs, T. J., Vondrak, R. R., Delory, G. T., & Halekas, J. S. 2007, GeoRL, 34, L14201

Farrell, W. M., Zimmerman, M. I., & Poppe, A. R. 2013, JGRA, 118, 20086

Feuerbacher, B., Anderegg, M., Fitton, B., et al. 1972, LPSC, 3, 2655

Freeman, J. W., & Ibrahim, M. 1975, Moon, 14, 103

Grün, E., Horanyi, M., & Sternovsky, Z. 2011, P&SS, 59, 1672

Hahn, J. M., Zook, H. A., Cooper, B., & Sunkara, B. 2002, Icar, 158, 360

Halekas, J. S., Bale, S. D., Mitchell, D. L., & Lin, R. P. 2005, JGR, 110, A07222

Halekas, J. S., Delory, G. T., Stubbs, J. K., & Farrell, W. M. 2009, JGR, 114, A05110

Halekas, J. S., Mitchell, R. P., Lin, L. L., et al. 2002, GeoRL, 29, 1436

Halekas, J. S., Poppe, A., Delory, G. T., Farrell, W. M., & Horanyi, M. 2012, EP&S, 64, 73

Jackson, T. L., Farrell, W. M., & Zimmerman, M. I. 2015, AdSpR, 55, 1710

Jordan, A. P., Stubbs, T., & Wilson, J. K. 2015b, JGRE, 120, 210

Jordan, A. P., Stubbs, T. J., Wilson, J. K., Schwadron, N. A., & Spence, H. E. 2015a, JGR, 120, 210

Manka, R. H. 1973, Photon and Particle Interactions with Surfaces in Space (New York: Springer), 347

Masuda, S. 1970, Advances in Static Electricity, Vol. 1 (Brussels: Auxilia, S.A.), 398

Masuda, S., Fujibayashi, K., Ishida, K., & Inaba, H. 1972, Electronic Engineering in Japan, 92, 9

Masuda, S., & Matsumoto, Y. 1973, Proc. 2nd Int. Conf. Static Electrification

Masuda, S., Washizu, M., & Kawabata, I. 1988, ITIA, 24, 217

McComas, D. J., Allegrini, F., Bochsler, P., et al. 2009, GeoRL, 36, L12104
Poppe, A. R., & Horanyi, M. 2010, JGRA, 115, 210JA015286
Porter, J. A., Townsend, L. W., Spence, H., et al. 2013, SpWea, 12, 329
Rennilson, J. J., & Criswell, D. R. 1974, Moon, 10, 2
Saito, Y. S., Yokoto, S., Tanaka, T., et al. 2008, GeoRL, 34, L24205
Schwadron, N. A., Spence, H. E., Wilson, J. K., et al. 2015, LPSC, 46, 2395
Singer, S. F., & Walker, E. H. 1962, Icar, 1, 7
Sorensen, T. C., & Spudis, P. D. 2005, JESS, 114, 645
Stubbs, T. J., Halekas, J. S., Farrell, W. M., & Vondrak, R. D. 2005, Workshop on Dust in Planetary Systems, Lunar and Planetary Institute 4070
Stubbs, T. J., Vondrak, R. R., & Farrell, W. M. 2006, AdSpR, 37, 59
Szalay, J. R., & Horanyi, M. 2015, GeoRL, 42, 5141
Tatom, F. B., Srepel, V., Johnson, R. D., et al. 1967, NASA Technical Report TR-792-7-207A, 3
Weiser, M., Barabash, S., Futaana, Y., et al. 2010, GeoRL, 37, L01502
Winslow, R. M., Jordan, A. P., Halekas, J. S., et al. 2015, LPSC, 46, 1261

Carlos I Calle and Karen L Aplin

Chapter 8

The Electrostatic Environment of Asteroids

Asteroids are small, irregular, airless, rocky solar system bodies ranging in size from meters to about 1000 km. Most asteroids orbit the Sun in the *Asteroid Belt*, a region between the orbits of Mars and Jupiter, extending from 2.2 AU to 3.3 AU (Figure 8.1). (1 Astronomical Unit, AU is the distance from the Earth to the Sun.) Their periods of orbital revolution around the Sun range from 3.3 to 6 yr. There are tens of thousands of these bodies but their total mass, at 8×10^{20} kg, is less than that of the Moon. The typical distance between asteroids larger than 1 km is of the order of several million kilometers.

There are three classes of asteroids: C-type asteroids (chondrite) are made up of clay and silicate rocks, are dark, and are among the oldest bodies in the solar system, which makes them highly scientifically important; S-type (stony) are made up of silicone materials and nickel–iron; and M-type (metallic) are made up of nickel–iron. More than eighty percent of the asteroids found at the outer edges of the asteroid belt are C-type asteroids. Their composition is similar to that of the carbonaceous chondrite meteorites that fall to Earth. In fact, these meteorites are probably pieces formed during asteroid collisions. S-type asteroids are found mostly at the inner edge of the asteroid belt and are the most abundant (Noble et al. 2010). The M-type asteroids are the least common and are found in the middle region of the asteroid belt.

Flybys of asteroids Gaspra and Ida by NASA's Galileo mission as well as spectral measurements of S-type asteroids, the most common type, indicate that *space weathering*, such as solar wind sputtering and micrometeorite bombardment, occurs on asteroids similarly to the Moon (Noble et al. 2010). However, on asteroids such as these, which are in the asteroid belt, space weathering processes are different. The solar wind flux is reduced at the larger distance from the Sun and micrometeorite impacts are slower but more frequent (Hortz & Schall 1981). Images taken by the NASA Near Earth Asteroid Rendezvous (NEAR)-Shoemaker spacecraft of the

doi:10.1088/2514-3433/ac878ech8

Figure 8.1. Image of the asteroid Eros, an S-type asteroid, taken by NASA's NEAR mission in 2000. (Courtesy of NASA.)

Figure 8.2. The dusty surface of Eros as photographed by NEAR-Shoemaker from a height of 250 m. The image is 12 m across. (Courtesy of NASA.)

asteroid 433 Eros, one of the largest near-Earth asteroids, show evidence of a layer of fine dust, with grain sizes under 50 μm in diameter (Figure 8.2).

8.1 The Asteroid Electrostatic Environment

Like the Moon, asteroids lack an atmosphere and are therefore fully exposed to the solar wind, solar UV radiation, and cosmic rays. Arguments similar to those for the Moon (discussed in Chapter 7) indicate that the interaction of the sunlit and night

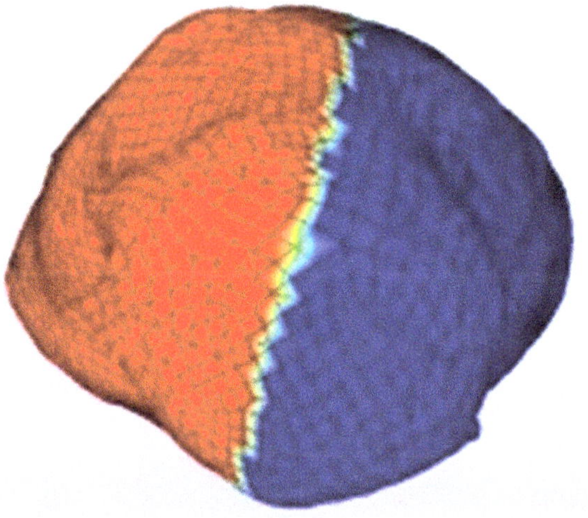

Figure 8.3. Potential at asteroid Itokawa, when the red (left) side is illuminated by the Sun and the right side is in darkness. The red color indicates +5 V and the blue −1000 V, with a transition zone at the terminator. These results are calculated with a 3D shape model for the asteroid's geometry, obtained by the Japanese Hayabusa sample return mission (Aplin et al. 2011).

sides of an asteroid with these charging currents would result in a Debye sheet that surrounds and shields the asteroid surface. On the sunlit side, photoelectron emission currents prevail, while on the night side, solar wind electron currents are dominant. Following the approach of Manka (1973) and of Mendis et al. (1981), Lee calculated the equilibrium surface potential at the sunlit surface of a main belt asteroid to be about +5 V (Lee 1996). On the night side surface, calculations show that the potential is about −200 V but can reach -10^3 V (Hortz & Schall 1981; Halekas et al. 2002). Along the terminator, where the photoelectron emission and solar wind electron currents balance, surface potentials change gradually from positive to negative, Figure 8.3.

8.2 Electrostatic Dust Transport

The dusty asteroid regolith may become charged by the different currents reaching the surface. The electrostatic fields present near the asteroid surface may transport these charged particles in different ways. In 1996, Lee predicted that electrostatic forces could levitate and transport dust on asteroids to "smooth, flat, and/or permanently shaded areas" (Lee 1996). In 2001, images from the NEAR-Shoemaker spacecraft showed smooth areas in craters 20–300 m in diameter that were interpreted as "ponds" (Veverka et al. 2001). The smoothness of the ponds as well as the colors of the pond material suggest that they are covered with dust grains 50 μm and under in diameter, consistent with Lee's electrostatic model (Robinson et al. 2001).

Dust particles on the sunlit side of an asteroid are expected to have a positive charge due to photoelectron emission. The electrostatic field in the photoelectron sheet, which points away from the asteroid surface, may levitate some of these positively charged particles. Depending on the particle's mass, the magnitude of its positive charge, its initial velocity, and the asteroid's gravity, the particle may stay inside the sheath or cross the layer. If the particle stays in the sheath, it may acquire a negative charge due to photoelectron absorption, which is greater than photoelectron emission. If the particle crosses the photoelectron sheath, it will emit photoelectrons and become positively charged (Colwell et al. 2005). It may then continue moving outward and leave the asteroid, fall back into the sheath, or even reach an equilibrium state, where the electrostatic force balances its weight.

Across the day/night boundary, called the *terminator*, or across boundaries between sunlit and shadow regions in craters, horizontal electric fields can form. In these cases, the electric fields will emanate from the positively charged sunlit side to the negatively charged shadow region. Charged dust can then be transported across these regions. A study of the particle dynamics on the asteroid Eros by Colwell et al. showed that electrostatic levitation of dust may provide enough dust to roughly account for the formation of the ponds observed by the NEAR-Shoemaker spacecraft (Colwell et al. 2005). The effect of the terminator on the potential of asteroid Itokawa is shown in Figure 8.3.

Shadows from visiting spacecraft can also disturb the local electrostatic environment at the asteroid. Calculations indicate that the asteroid's potential could be rapidly changed by the arrival of a spacecraft, Figure 8.4. Before the spacecraft arrives, the dayside potential is 5 V, consistent with the theory described in Section 8.1 above, but the effect of the spacecraft's shadow as it approaches to land reduces it to about −4 V at 3 m away. In Figure 8.4 the spacecraft was simply represented by a square of side 1 m, however in reality it could have significantly larger solar panels, increasing the shadowed area. Since the principle of a sample return mission is that a pristine sample from the landing site is collected and returned, the possibility that the spacecraft itself could produce charged dust transport across the shadow boundary should be investigated further (Aplin et al. 2014).

8.3 Cohesive Forces in Asteroids

Analyses of the distribution of asteroid spin rates versus size have shown that asteroids larger than a few hundred meters in diameter are in general loosely bound aggregates with minimal tensile strength (so-called *rubble pile* asteroids; Pravec & Harris 2000). Smaller asteroids, with diameters under 100 m, are monolithic. The JAXA Hayabusa mission to asteroid Itokawa, which has a mean radius of 162 km, successfully landed a spacecraft on the asteroid in 2005 November and collected samples. Measurements of those samples and estimates of the sizes of rocks showed that the size distribution of the regolith components ranges from under 100 μm to millimeters in diameter for granular material and from millimeters to 40 m for rocks and boulders (Tsuchiyama et al. 2011). Observations of Main Belt asteroids have shown that regolith dust sizes extend down to about 10 μm (Jewitt et al. 2010).

Some studies have looked at the nature of the cohesive forces keeping the aggregate material of an asteroid together. Measurements of the rotation period and

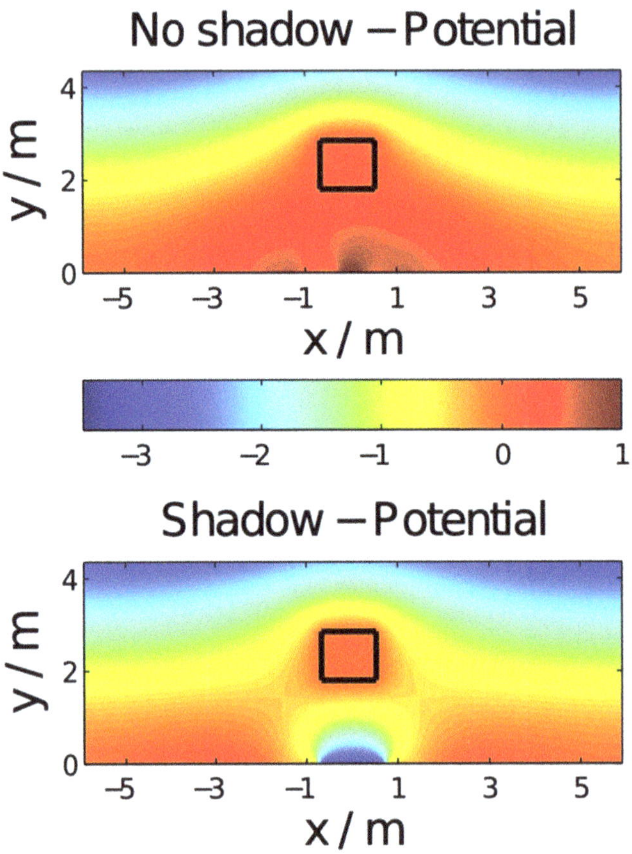

Figure 8.4. Effect of shadowing from a 1 m^2 spacecraft on the local potential at 3 m from the surface of an asteroid. The top plot shows local modification of the potential by the spacecraft without the shadow effect, and the bottom plot includes the shadow. In this example the Sun is directly overhead in both plots. The voltage color scale is between the top and bottom plots. Reproduced from Aplin et al. (2014) with permission from Elsevier.

bulk density of the near-Earth asteroid 1950 DA indicate that it is rotating faster than gravity and friction would allow (Rozitis et al. 2014). This asteroid has a mean diameter of 1.3 km and a rotation period of 2.1216 h, faster than the estimated critical spin limit of 2.2 h for a coherent or monolith asteroid (Harris 1996). The measured rotation period clearly requires the presence of cohesive forces. The possible forces are electrostatic and van der Waals forces between grains. The van der Waals force between two neutral objects is due to the interaction between induced dipoles that result from polarizable atoms in the objects. The van der Waals force decreases rapidly with separation between objects, becoming negligible at distances over 30 nm or so. When electrostatic forces are also present, as is usually the case, these two interactions have about the same strength at this distance of 30 nm. At greater separations, the electrostatic force prevails. At 10 nm, van der Waals has about ten times the strength of the electrostatic force (Gady et al. 1996). Van der

Waals forces are significant only for particles with diameters under 100 μm and predominate for particles under 10 μm (Casterllanos 2005). Thus, besides gravity and friction, the cohesion forces holding together rubber pile asteroids may be van der Waals forces between fine grains that may act as the glue that holds rocks and boulders together along with electrostatic forces acting on the larger grains and rocks.

For the 1950 DA asteroid, Rozitis et al. found that a minimum cohesive strength of 64 Pa is required to keep the asteroid intact, which is in the 3–300 Pa range determined by numerical simulations and in rough agreement with the value of 100 Pa found for lunar regolith (Jewitt et al. 2010; Mitchell et al. 1974; Sanchez & Scheeres 2014). Sanchez and Scheeres ran a simulation of a boulder immersed in regolith using a 1 m boulder half-immersed in a hemispherical container filled with particles between 2 and 3 cm in diameter (Sanchez & Scheeres 2014). Using a value of 1 mg for the gravitational field and actual asteroid spin data, they obtain a value of 25 kPa for the cohesive strength of a rubble pile asteroid. This value is consistent with rubble piles containing dust particles with diameters under 10 μm acting as glue for larger boulders.

NASA's OSIRIS-REx (Origins, Spectral Interpretation, Resource Identification, Security–Regolith Explorer) sample return mission from asteroid Bennu (Lauretta et al. 2017) collected 250 g of asteroid material that will be returned to Earth in 2023 for laboratory analysis. During the sample collection maneuver, the spacecraft started to sink into the surface before it fired its thrusters with a jet of nitrogen gas that propelled it away from Bennu. The jet of nitrogen gas levitated material from the surface that was collected in a chamber on the spacecraft. An initial analysis of the images as well as spectral and accelerometer data taken during this touch and go maneuver indicates that the surface particles are loosely bound and that the interparticle cohesion forces near the surface are very weak (Walsh et al. 2022). More detailed laboratory analysis for the collected samples from OSIRIS-REx and other planned missions should increase our understanding of the role of cohesive and electrostatic forces at asteroids.

References

Aplin, K. L., Bowles, N. E., Urbak, E., Keane, D., & Sawyer, E. C. 2011, JPhCS, 301, 012008

Aplin, K. L., Macfaden, A. J., & Bowles, N. E. 2014, P&SS, 99, 103

Casterllanos, A. 2005, AdPhy, 54, 263

Colwell, J. E., Gulbis, A. A. S., Horányi, M., & Robertson, S. 2005, Icar, 175, 159

Gady, B., Schleef, D., Reifenberger, R., Rimai, D., & Demejo, L. P. 1996, PhRvB, 53, 8065

Halekas, J. S., Mitchell, D. L., Lin, R. P., et al. 2002, GeoRL, 29, 1436

Harris, A. W. 1996, LPSC, 27, 493

Hortz, F., & Schall, R. B. 1981, Icar, 46, 337

Jewitt, D., Weaver, H., Agarwal, J., Mutchler, M., & Drahus, M. 2010, Natur, 467, 817

Lauretta, D. S., Balram-Knutson, S. S., Beshore, E., et al. 2017, SSRv, 212, 925

Lee, P. 1996, Icar, 124, 181

Manka, R. H. 1973, Photon and Particle Interactions with Surfaces in Space, ed. R. L. Grard (Dordrecht: Reidel), 347

Mendis, D. A., Hill, J. R., Houpis, H. L. F., & Whipple, E. C. 1981, ApJ, 249, 787
Mitchell, J. K., Houston, W. N., Carrier, W. D., & Costes, N. C. 1974, Apollo Soil Mechanics Experiment S-200 Final Report, NASA
Noble, S. K., Keller, L. P., & Pieters, C. M. 2010, M&PS, 45, 2007
Pravec, P., & Harris, A. W. 2000, Icar, 148, 12
Robinson, M. S., Thomas, P. C., Veverka, J., Murchie, S., & Carcich, B. 2001, Natur, 413, 396
Rozitis, B., MacLennan, E., & Emery, J. P. 2014, Natur, 512, 174
Sanchez, P., & Scheeres, D. J. 2014, M&PS, 49, 788
Tsuchiyama, A., Uesugi, M., Matsushima, T., et al. 2011, Sci, 333, 1125
Veverka, J., Thomas, P. C., Robinson, M., et al. 2001, Sci, 292, 484
Walsh, K. J., Ballouz, R.-L., Jawin, E. R., et al. 2022, SciA, 8, eabm6229

Carlos I Calle and Karen L Aplin

Chapter 9

The Martian Electrostatic Environment

9.1 The Martian Atmosphere

The Martian atmosphere is composed of 95.32% carbon dioxide, 2.7% nitrogen, 1.6% argon, 0.13% oxygen, 0.08% carbon monoxide, and traces of water (0.02%), nitrogen oxide, neon, krypton, and xenon. The average atmospheric pressure near the surface of the planet is 6.36 mb (1 millibar = 4.77 Torr), ranging from 4.0 to 8.7 mb, which is about 0.6% of the Earth's atmospheric pressure near the surface. The density of the Martian atmosphere decreases by a factor of $1/e$ relative to the density at the surface at an altitude of 11.1 km. The altitude at which the density and pressure of an atmosphere decreases to this factor is known as the *scale height* of the atmosphere. The scale height of the Earth's atmosphere is 8.6 km.

Like the Earth, the atmosphere of Mars is divided into distinct layers (Figure 9.1). Closest to the surface is the troposphere, where convection takes place. The height of the troposphere changes with surface temperature. During the day in the Martian summer, surface temperatures can reach 300 K and the height of the troposphere can reach 30 km. During night time in the polar regions, where the temperature is lowest, the troposphere disappears altogether. Above the troposphere is the stratosphere, where the temperature stays fairly constant up to about 140 km. The stratosphere and troposphere are separated by a region known as the *tropopause*. At higher altitudes, the temperature rises steadily.

Water ice clouds occasionally form in the troposphere, usually around mountains. Higher up, in the stratosphere, carbon dioxide condenses and forms clouds and haze.

Dust is uploaded into the atmosphere from the planet's surface by a mechanism that is not completely understood. The layer of dust and sand that covers the planet has been somewhat homogenized by global dust storms that take place every few Martian years. Dust devils up to 10 km high are relatively common on Mars. Some of the landing locations for the Mars Exploration Rovers (MERs) have detected

doi:10.1088/2514-3433/ac878ech9

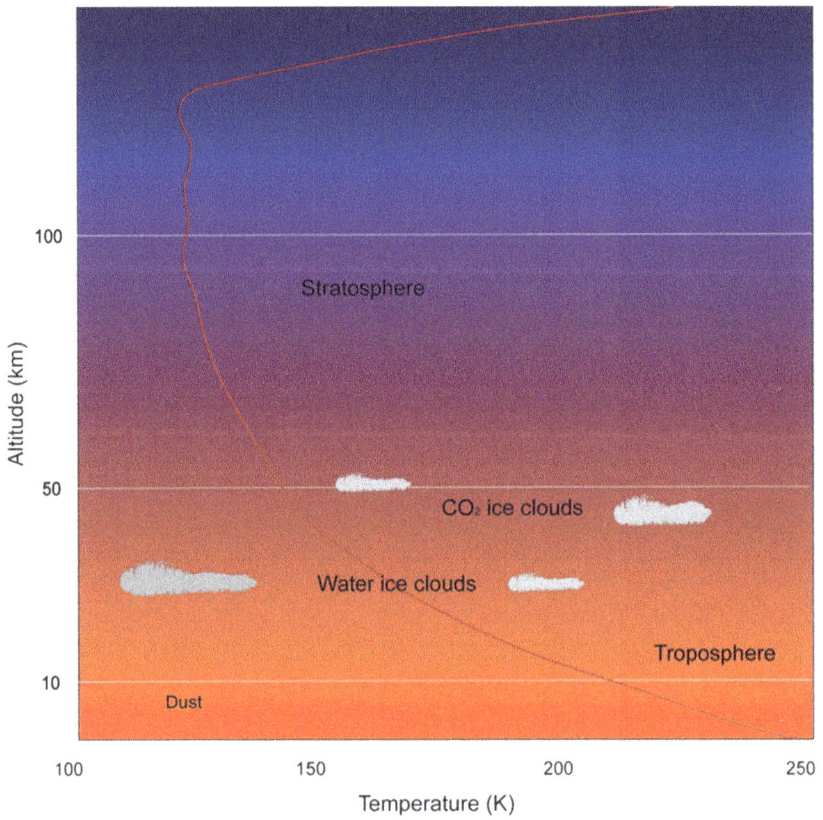

Figure 9.1. Temperature variation with altitude in the Martian atmosphere.

dust devils in their vicinity almost daily. It is perhaps this dust storm activity which generates the mechanism that raises dust into the tenuous Martian atmosphere. It has been proposed that saltation in the presence of electric fields may uplift the dust (Harrison et al. 2016).

Landis et al., analyzing data from the Microscopic Imager on the NASA MERs, have found that the dust content of the atmosphere has a three-component particle distribution: *atmospheric dust* suspended for long periods of time, with diameters from 2 to 4 μm; *settled dust* uplifted into the atmosphere by dust devils and wind storms, with particle diameters of 10 μm and under; and *saltating particles*, with diameters over 80 μm (Landis et al. 2006). The Microscopic Imager also measured the average diameter of particles on the surface of Mars to be 220 μm.

Although the concentration of dust in the Martian atmosphere has never been measured directly, these values can be obtained from measurements of the opacity of the atmosphere that have been taken from landers. The opacity of the atmosphere is measured in terms of the *optical depth* τ, which is a measure of the transmission of radiation through the atmosphere. Optical depth is given by the logarithm of the ratio of transmitted to incident radiant power through the atmosphere.

Typical values of the optical depths in the Martian atmosphere during non-dust storm conditions range from 0.2 to 1 (Lemmon et al. 2014). Optical depths during local dust storm conditions are in the range from 1 to 6. Figure 9.2 shows optical depths measured by the MERs Spirit and Opportunity during 5 years of their mission.

Using the MER optical depth data, we can calculate the expected atmospheric dust particle density for different conditions. The number of particles per unit volume as a function of height z can be approximated from (Haberle et al. 1993)

$$N \sim N_0 \tau e^{-\frac{z}{H}}$$

where N_0 is the number per unit volume at the surface for an optical depth of 1 and H is the scale height, which has an average value of 11.1 km. For relatively clear atmospheric conditions, with the optical depth τ from 0.2 to 1, the average number of dust particles in the atmosphere near the ground ($z = 0$) ranges from about 5–24 particles cm^{-3}. For dust storm conditions, using $\tau = 6$, the expected particle density is about 140 particles cm^{-3}.

By comparison, a typical terrestrial indoors environment is considered similar to a class 100,000 clean room, a classification used in the United States by the electronics, pharmaceutical, and medical industries. A class 100,000 clean room contains 100,000 particles 0.5 μm and larger in diameter per cubic foot of air. This value is equal to 3.5 particles cm^{-3}, which falls at the low end of the range of the atmospheric particle density during non-dust storm conditions on Mars. However, the Martian atmosphere has a density of 0.020 kg m^{-3} near the surface, which is about 1.6% of the density of the terrestrial atmosphere near the surface. If we were to pump Martian atmospheric gas into a chamber and increase its density to match that of the Earth's atmosphere, the particle concentration would increase from an average of about 11 (taking the middle of the range for calm conditions) to about 670 particles cm^{-3}.

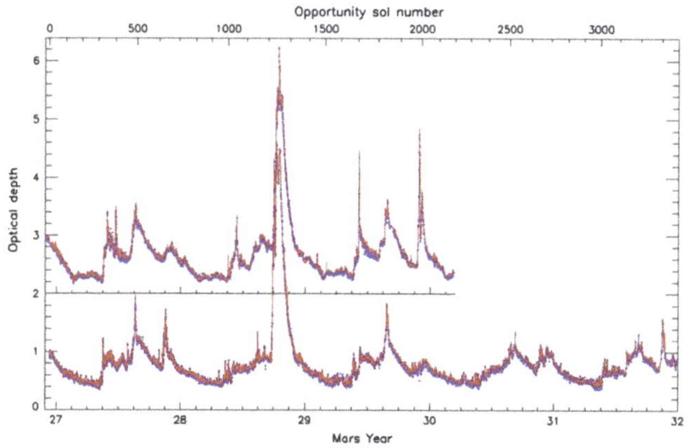

Figure 9.2. MERs' optical depth for its first 1200 sols showing Spirit (bottom) and Opportunity (top) data. Opportunity's data is offset by 2 for clarity. (Reproduced from Lemmon et al. 2014, with permission from Elsevier.)

NASA's Mars human exploration program relies on the utilization of local resources for astronaut sustainability. A plant that would pump Martian atmospheric gas into a chemical reactor chamber for oxygen, methane, and water extraction would contain enough dust to clog up filters and potentially damage reactors. Methods to use electrostatic forces to extract dust particles from these planned intakes are being studied (Calle et al. 2011).

9.2 Electrical Breakdown in the Martian Atmosphere

At about 0.6% of the Earth's atmospheric pressure, breakdown in the Martian atmosphere occurs at a much lower potential than on Earth. Since the Martian atmosphere is mostly made up of carbon dioxide, a quick glance at Figure 9.3 indicates that with a 1 mm gap, breakdown in CO_2 will take place at about 450 V. Near the Earth's surface, breakdown with this gap requires 4.3 kV. Experimental data comparing breakdown in CO_2 and in a premixed gas that emulates the Martian atmosphere shows that the breakdown potentials are nearly identical (Figure 9.3; Calle et al. 2011). These experiments were carried out in a vacuum chamber at a pressure of 9 mb. This data indicates that the CO_2 dominates the Paschen breakdown and that the additional gases modify the breakdown potential by an average of only 15 V.

9.3 Electrostatic Charge and Size of Martian Atmospheric Dust Particles

Triboelectrification during dust storms or devils produces positively and negatively charged grains. On Mars, convective instabilities in the atmosphere should stratify the charged dust grains, with lighter grains being lifted to higher altitudes than more massive grains. Since, in dust, smaller particles tend to charge negatively and larger particles charge positively (Forward et al. 2009), a macroscopic dipole moment is formed in the atmosphere that can produce an electrical discharge (Farrell et al. 1999).

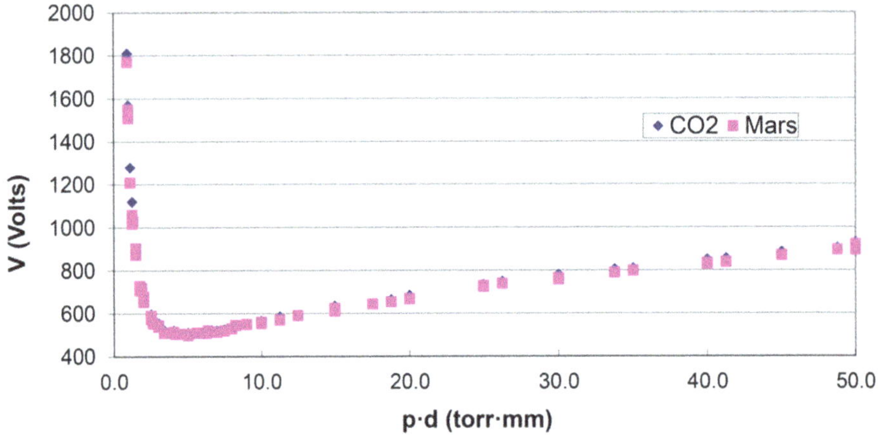

Figure 9.3. Paschen breakdown potentials versus pressure–distance for a Martian gas mixture (red squares) and for CO_2 (blue triangles) (Calle et al. 2011).

Several groups of researchers have demonstrated experimentally in simulated Martian atmospheres that this type of dust vertical motion can generate electric fields strong enough for electrical discharges to occur (Harrison et al. 2016).

Numerical models of electrification during Martian dust storms (Farrell et al. 2003) and devils (Zhai et al. 2006) predict that these electric fields should have strengths up to the theoretical breakdown potential of carbon dioxide at the low atmospheric pressure near the surface of Mars. Combined with experimental values of electron concentration in the Martian atmosphere, the models yield values of the electrical conductivity of the atmosphere that are several orders of magnitude higher than the values for the terrestrial atmosphere (Harrison et al. 2016). If these estimates are correct, charge dissipation in the Martian atmosphere away from dust storms would take place in seconds rather than the minutes it takes for similar processes to act at Earth. In dust devils or storms the dust attaches to ions and electrons to reduce the conductivity and increase the time charge takes to dissipate. The discharge mechanism, however, remains unknown. Whether this discharge takes place violently, through lightning, or more gently, as in corona glow, is not known. No measurements of lightning or glow discharges in the Martian atmosphere have ever been made. However, there is some experimental evidence for glow discharge in laboratory experiments. Eden and Vonnegut placed sand particles in a container with carbon dioxide at pressures in the range of the Martian atmospheric pressure and observed a glow as well as filamentary electrical discharges when the container was shaken (Eden & Vonnegut 1973). More recently, a NASA laboratory conducted similar experiments where a visible glow was observed; the discharges altered several organics known to exist on Mars (Hintze et al. 2010). Glow discharges would be relatively faint and difficult to detect from orbit if they take place in thick dust storms. In contrast with these findings and earlier results, one charging model showed that electric fields cannot reach levels up to breakdown because of charge dissipation in the saltation layer (Kok & Renno 2009).

Searches for evidence of electrostatic discharges in the Martian atmosphere have been made with instrumentation aboard orbiting spacecraft. In 2009, Ruf and collaborators claimed that they had detected non-thermal electromagnetic emissions during a dust storm. Analyses of the modes of these emissions were interpreted to be Schumann resonances from lightning discharges exciting the atmosphere, like on Earth. However, a subsequent observation in the same electromagnetic region by Anderson and colleagues found no evidence of Schumann resonances during a period that included dust storms. Detailed studies of over five years of observations by the Mars Advanced Radar for Subsurface and Ionosphere Sounding (MARSIS) yielded no evidence of high-frequency radio emissions that would indicate the presence of electrical discharges. This material is discussed in more detail in Harrison et al. (2016).

A key outstanding question related to the presence of lightning and glow discharges in the Martian atmosphere is the rate of charge dissipation. The atmosphere probably contains free electrons, making it relatively conductive, which would limit charge build-up. However, in dust storms the conductivity will be lower due to electrons sticking to the dust particles, and charge is also generated from dust–dust interactions, which would make discharges more likely.

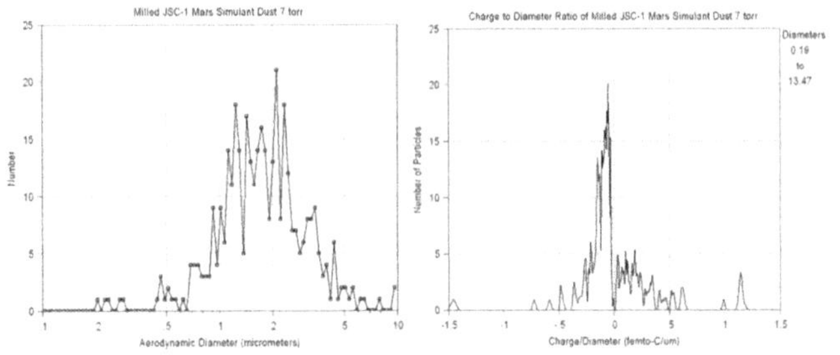

Figure 9.4. Aerodynamic diameter (left) and electrostatic charge density (right) of simulated Martian dust particles measured in a low-pressure carbon dioxide atmosphere. The diameters of the particles used are about half of the observed diameters on Mars (Buhler et al. 2008). (Credit: Society for Imaging Science and Technology.)

Experiments to measure the charge and aerodynamic diameter of dust particles in a simulated Martian atmospheric environment were carried out in the NASA Electrostatics and Surface Physics laboratory in collaboration with the University of Arkansas at Little Rock (Buhler et al. 2008). Electrostatic charge and aerodynamic diameters for simulated Martian dust particles were measured in a carbon dioxide atmosphere at a pressure of 6 mbar (Figure 9.4). The average particle diameter used in these experiments is smaller than the actual particle diameters derived from optical data taken from spacecraft in orbit around Mars and with lander instruments. The dust cross section weighted mean radius at Gusev crater where Spirit was operating was 1.47 ± 0.21 μm, and 1.52 ± 0.18 μm at Meridiani crater, Opportunity's home (Lemmon et al. 2004). The experiments showed the feasibility of an instrument that could be deployed in a future mission to directly measure both particle diameter and electrostatic charge.

References

Buhler, C. R., Mazumder, M. K., Calle, C. I., et al. 2008, Proc. Pan Pacific Imaging Conf.

Calle, C. I., Thompson, S. M., Cox, N. D., et al. 2011, JPhCS, 327, 012048

Eden, H. F., & Vonnegut, B. 1973, Sci, 180, 962

Farrell, W. M., Kaiser, M. L., Desch, M. D., et al. 1999, JGR, 104, 3795

Farrell, W. M., Delory, G. T., Cummer, S. A., & Marshall, J. R. 2003, GeoRL, 30, 2050

Forward, K. M., Lacks, D. J., & Sankaran, R. M. 2009, GeoRL, 36, L13201

Haberle, R. M., McKay, C. P., Pollack, O. E., et al. 1993, Atmospheric Effects on the Utility of Solar Power for Mars, in Resources of Near-Earth Space, ed. J. Lewis, & M. S. Matthews (Tucson, AZ: Univ. Arizona Press)

Harrison, R. G., Barth, E., Esposito, F., et al. 2016, SSRv, 203, 299

Hintze, P. E., Buhler, C. R., Schuerger, A. C., Calle, L. M., & Calle, C. I. 2010, Icar, 208, 749

Kok, J. F., & Renno, N. O. 2009, GeoRL, 36, L05202

Landis, G. A., Herkenhoff, K., Greeley, R., et al. 2006, LPS, 37, 1932

Lemmon, M. T., Wolff, M. J., Smith, M. D., et al. 2004, Sci, 306, 1753

Lemmon, M. T., Wolff, M. J., Bell, J. F. III, et al. 2014, Icar, 251, 96

Zhai, Y., Cummer, S. A., & Farrell, W. M. 2006, JGRE, 111, E06016

Chapter 10

The Electrostatic Environments of Venus and Mercury

10.1 Electrical Phenomena in the Venusian Atmosphere

Venus is in many ways very similar to the Earth. The two planets have about the same size, density, and chemical composition. However, the Venusian atmosphere is made up mostly of carbon dioxide (96.5% CO_2 and 3.5% nitrogen) with a density about 90 times that of the Earth and a temperature near the surface that reaches 730 K. This high temperature is due to the much denser atmosphere that traps 99% of the infrared radiation reradiated from the surface. This greenhouse effect is the most powerful in the solar system.

The lower atmosphere is mainly clear up to about 32 km. Above that, a thin haze of dust and aerosols reaches up to about 47 km, where the 20 km thick cloud layer that surrounds the planet starts. The lower part of this layer contains large solid particles that were discovered by the Pioneer Venus mission and that are of unknown origin. The outer part of the layer is made up of small sulfuric acid droplets. These droplets could in principle acquire large electric charges and contribute to the generation of lightning. Although there are no direct measurements of Venus cloud charging, observations from terrestrial telescopes as well as with instrumentation on NASA's Pioneer Venus and ESA's Venus Express provided evidence for lightning in the Venusian atmosphere. Chris Russell and collaborators consider this evidence to be overwhelming (Russell et al. 2011). However, some theoreticians are not convinced. Mary Michael et al., for example, conclude that the expected electrical conductivity of the atmosphere and the expected abundance of aerosols do not support lightning on Venus (Michael et al. 2009). There is also a possibility of upper atmospheric discharges on Venus, which are lightning from the tops of the clouds upwards. These are a family of discharges known as *transient*

doi:10.1088/2514-3433/ac878ech10

luminous events (TLEs) including exotically named phenomena such as sprites and elves. These are often faint, but may be easier to spot from an orbiting spacecraft than lightning within the thick cloud layer (Lorenz 2018).

Unlike Earth, Venus lacks an intrinsic magnetic field. Thus the solar wind particles interact directly with the upper atmosphere, since there is no magnetosphere to deflect them. UV radiation from the Sun ionizes atoms in the upper atmosphere, creating a highly conductive ionosphere. This ionosphere deflects the solar wind around the planet, forming a bow shock and a magnetotail (Zhang et al. 2006). The interaction of the solar wind with the top of the ionosphere produces a complex electrical environment that causes the ejection into space of charged particles in the magnetotail. Venus Express found that this environment also sends solar wind particles deep into the upper atmosphere, where they collide with the carbon dioxide molecules, producing the faint flashes of light that had been observed before. Data from the Venus Express magnetometer and the low-energy particle detector showed a plasma flow toward the planet that was 3400 km wide that lasted for 94 s (Zhang et al. 2012).

Venus Express ended its mission in 2014. On 2015 December 6, Akatsuki, the Venus Climate Orbiter from the Japanese Space Agency, went into orbit around Venus. The craft is studying the atmosphere and its cloud layer, and is looking for additional evidence for lightning with its Lightning and Airglow Camera. NASA is now planning two new missions to Venus to understand the formation and evolution of its atmosphere as well as its geologic history. The DAVINCI+ (Deep Atmosphere Venus Investigation of Noble gases, Chemistry, and Imaging) mission will consist of a sphere that will descend through the dense Venusian atmosphere to make precision measurements of noble gases and other gases in the atmosphere (Figure 10.1). The second mission, called VERITAS (Venus Emissivity, Radio Science, InSAR, Topography, and Spectroscopy), will map the planet's surface to understand why Venus evolved differently from Earth. (The mission will also carry the Deep Space Atomic Clock 2, built by JPL, to eventually support autonomous spacecraft maneuvers and to enhance radio measurements.) The European Space Agency is also planning the Envision orbiter to study the evolution of Venus. These observations may provide stronger evidence for lightning on Venus.

10.2 The Electrostatic Environment of Mercury

Mercury, the planet closest to the Sun, has an extremely thin atmosphere with a surface pressure of about 5×10^{-15} bar, which is essentially a vacuum. Mercury is the smallest planet in the solar system, with a diameter of 4879 km, it is only slightly larger than the Moon (Figure 10.2). Like the Moon, Mercury is exposed to the solar wind, cosmic rays, and the Sun's UV radiation. The surface of the planet acquires an electric potential that balances all the current fluxes reaching the surface. There are multiple currents, ranging from the protons and electrons in the solar wind; alpha particles, protons, and electrons in cosmic rays; photoelectrons; to secondary electrons and GCRs partially reflected from the surface. Of those, only the solar

Figure 10.1. Conceptual image of Venus DAVINCI+atmospheric probe. (Courtesy of NASA.)

wind protons and electrons, the photoelectrons, and the secondary electrons contribute most of the currents (Stubbs et al. 2005).

Mercury, like the Moon and all other airless bodies in the solar system, is a charged body in a plasma. A photoelectron sheath forms on the side facing the Sun and a Debye sheath forms on the dark side. The solar wind flux is larger around Mercury, since it is much closer to the Sun, than it is around the Moon. Potentials on the surface should still be around +10 V on the day side of Mercury and in the negative hundreds of volts in the plasma wake of the night side, reaching higher values during solar storms.

A similar situation is expected around all moons without appreciable atmospheres. Titan and other moons with atmospheres should have different electrostatic environments dominated by their weather. Current and past exploration missions have yielded no evidence of atmospheric electricity on those bodies. But, as in the case of Venus, lightning is not easy to detect from orbit. Future missions may provide a better picture of the electrostatic environments of these bodies.

Figure 10.2. Mercury from NASA's MESSENGER spacecraft, which studied the planet from 2008 to 2015. (Courtesy of NASA.)

References

Lorenz, R. D. 2018, PEPS, 5, 34

Michael, M., Tripathi, S. N., Borucki, W. J., & Whitten, R. C. 2009, JGR, 114, E4

Russell, C. T., Strangeway, R. J., Daniels, J. T. M., Zhang, T. L., & Wei, H. Y. 2011, P&SS, 59, 956

Stubbs, T. J., Halekas, J. S., Farrell, W. M., & Vondrak, R. D. 2005, Workshop on Dust in Planetary Systems (Lunar and Planetary Institute), 4070

Zhang, T. L., Baumjohann, W., Delva, M., et al. 2006, P&SS, 54, 1336

Zhang, T. L., Lu, Q. M., Baumjohann, W., et al. 2012, Sci, 336, 567

Carlos I Calle and Karen L Aplin

Chapter 11

The Electrostatic Environment of Jupiter

11.1 The Electrostatic and Magnetic Environments of Jupiter

Jupiter, the largest planet in the solar system, has an atmosphere composed mainly of hydrogen (89.8% by volume) and helium (10.2%) gases, similar to that of the Sun. Its clouds of ammonia crystals, ammonium hydrosulfide crystals, and frozen water form the red, white, and brown bands visible from Earth even with small telescopes. Slightly deeper within the atmosphere, there are liquid water and water-ice clouds like those on Earth. The Great Red Spot is a giant storm larger than the Earth that has existed for hundreds of years.

At the lower layers of the atmosphere, the high pressure compresses the hydrogen gas into a liquid, forming an ocean around the entire planet. Deeper inside Jupiter, the huge atmospheric pressure dissociates hydrogen molecules in this vast ocean and produces an ionized plasma, making the liquid hydrogen electrically conductive (Pierleoni et al. 2016). The rapid rotation of the planet creates electrical currents that generate Jupiter's strong magnetic field (Figure 11.1). This field engulfs the planet with an extensive magnetosphere that reaches out beyond the orbit of many of its moons[1] on the side facing the Sun, extending into a long tail that reaches beyond the orbit of Saturn. Jupiter's rotationally driven magnetosphere derives its energy from the planet's rotation rather than from the solar wind, like the Earth's magnetosphere (Khurana et al. 2004). Jupiter's magnetic field traps large quantities of charged particles in belts, analogous to Earth's Van Allen Belts. These particles are a hazard for visiting spacecraft, which need special shielding, as discussed in Chapter 3. Due to the planet's rapid rotation, some of these charged particles are ejected into an immense current sheet 2–4 Jupiter radii in thickness that lies almost precisely in the

[1] At the time of writing Jupiter has 53 officially named moons according to the International Astronomical Union, and at least 79 in total.

doi:10.1088/2514-3433/ac878ech11

Figure 11.1. Artist's rendition of Jupiter's magnetic field lines. (Courtesy of NASA.)

magnetic equatorial plane (Goertz 1976). The main source of plasma particles in Jupiter's magnetosphere is its moon Io, which contributes sulfur and oxygen ions. A second plasma source in Jupiter's magnetosphere is the solar wind, which contributes protons and electrons. Although the mass density of the solar wind is much lower than the density of contributions from Io, the number density of protons in the solar wind reaching Jupiter's magnetosphere is comparable to that of Io (Hill et al. 1983). Protons (H^+) and molecular hydrogen ions (H^+) from Jupiter's ionosphere are the third plasma source in Jupiter's magnetosphere. Compared to the other two, the ionosphere is not a major source of plasma. Finally, sputtering of the surfaces of Europa, Ganymede, and Callisto, the three icy moons of Jupiter, contributes a small amount of plasma to the magnetosphere (Cooper et al. 2001).

High-energy plasma particles can penetrate Jupiter's atmosphere near the magnetic poles and excite hydrogen and helium gases which in turn emit energy in the visible part of the spectrum, producing the largest and most energetic auroras in the solar system. In 2016, astronomers started an observation program to observe and measure auroras on Jupiter with the Hubble Space Telescope Imaging Spectrograph (Figure 11.2), and these measurements have been continued by the NASA Juno mission described below.

11.2 Lightning on Jupiter

NASA's Voyager 1, Galileo, Cassini and New Horizons missions obtained optical images of lightning on Jupiter. Analysis of the images indicates that lightning occurs at a rate of 4×10^{-3} km^{-2} s^{-1}, much lower than the rate on Earth of 6 km^{-2} s^{-1} (Uman 1987). However, lightning on Jupiter appears to be far more energetic than on Earth. The average energy per bolt of lightning on Jupiter is about 10^{12} J, much larger than that on Earth, which has an average energy per bolt of 10^9 J.

Besides visual and photographic detection of lightning bolts, lightning is detected on Earth with radio receivers that measure the radio frequency (RF) radiation emitted by thunderstorms. For terrestrial lightning this RF peaks at a frequency of 10 kHz, in the very low frequency (VLF) region of the spectrum (3–30 kHz) and extends to even lower frequencies, known as the extremely low frequency (ELF) region. As these ELF

Figure 11.2. NASA's Hubble Space Telescope image of an aurora near Jupiter's magnetic pole. (Courtesy of NASA.)

frequencies are reflected by the Earth's ionosphere and remain in the atmosphere, these radio waves are called *atmospherics*, which has been shortened to *sferics*. Higher-frequency VLF waves can penetrate the ionosphere and travel along the Earth's magnetic field lines. Because they produce tones that appear to glide up and down in pitch when played through a loudspeaker, they are called *whistlers*. As these whistler waves interact with energetic electrons in the Earth's magnetosphere, they generate chorus and hiss waves (see Chapter 5; Shawhan 1979).

Voyager 2 detected whistlers in Jupiter's atmosphere with its Plasma Wave Subsystem (PWS) instrument (Gurnett et al. 1979). Analysis of this data showed the lightning rate to be 4×10^{-3} km^{-2} s^{-1}, which agrees with the rate obtained from optical detection. The Galileo probe detected RF emissions from lightning as it descended into the Jovian atmosphere (Rinnert et al. 1998), but lightning-generated radio signals were not detected outside the ionosphere (Desch et al. 2002). One suggested explanation for this was strong absorption by the ionosphere. William Farrell at the NASA Goddard Space Flight Center also proposed that since lightning discharges on Jupiter appear to be a hundred times slower than on Earth, the frequency of most of the emitted radiation would lie below the cutoff frequency required to penetrate the planet's ionosphere and were therefore not detected by any spacecraft (Farrell 2000).

Understanding of Jupiter lightning has developed rapidly since the arrival of NASA's latest mission, Juno (Figure 11.3). Juno arrived at Jupiter in July of 2016 and was originally scheduled to orbit the planet until 2021. The mission is now in an

Figure 11.3. Artist's concept of the Juno spacecraft as it arrives at Jupiter. (Courtesy of NASA.)

extended phase until 2025 or the end of life of the spacecraft. At the end of the mission Juno will deorbit and plunge into the atmosphere to prevent any contamination of Jupiter's satellites. Some of Jupiter's moons are thought capable of harboring life, so it is important to protect them from any terrestrial contamination (Greenberg 2008).

Among Juno's mission goals are to explore and study its magnetosphere near the poles to obtain new information on how the planet's strong magnetic field affects its atmosphere. Juno carries five sets of instruments. The magnetometers study Jupiter's deep structure by mapping the planet's magnetic fields; a microwave radiometer (MWR) probes Jupiter's deep atmosphere to measure the amount of water; the Jupiter Energetic-particle Detector Instrument (JEDI), the Jovian Auroral Distributions Experiment (JADE), and the Plasma Waves Instrument (WAVES) sample electric fields, plasma waves, and charged particles around Jupiter to determine how the magnetic field links to the atmosphere; the UVS and JIRAM UV and infrared cameras provide images of the electrical activity in the atmosphere and on the auroras; and the JunoCam takes close-up photographs of Jupiter.

Juno's instruments have made major advances in understanding three aspects of Jupiter lightning. First, Juno was able to detect the characteristic radio emissions from lightning outside Jupiter's ionosphere for the first time. The MWR detected impulses in the ultra-high frequency (UHF: 300–3000 MHz) range at the same time as WAVES detected whistlers (Aplin et al. 2020). These radio signals were corroborated by simultaneous optical detections (Imai et al. 2020). The UHF signals were unexpected because lightning signals decrease in intensity with increasing frequency, meaning that terrestrial lightning, which is weaker to start with, produces almost no signal in the UHF range. Searching the UHF region of the radio spectrum can be considered a new way to detect Jupiter lightning. Second, Juno made the first observations of an extraterrestrial transient luminous event (TLE;

Giles et al. 2020), a type of upward-moving discharge (see Chapter 4). TLEs have been regularly photographed on Earth since the 1990s and predicted for other planets (Yair et al. 2009), but not observed until Juno.

Finally, a new type of lightning, called shallow lightning, was identified near the top of the clouds, from images from one of Juno's navigation cameras combined with MWR data. Jovian lightning normally originates from parts of the atmosphere containing clouds resembling terrestrial thunderclouds. These clouds seem to generate lightning in the same way as on Earth, by interactions between liquid water, ice and a type of soft hail called *graupel*. However, the clouds producing the shallow lightning are made of ammonia, water-ice and "mush balls", an ammonia–water mixture with a water-ice crust. Lightning originating from ammonia clouds is an exciting finding, as it demonstrates that Jupiter hosts the first type of lightning in the solar system generated via a mechanism not already seen on Earth (Becker et al. 2020). Because lightning is generated from collisions between cloud particles in turbulent motion, understanding Jupiter lightning is key to understanding Jupiter's weather systems in general. It seems likely that the Juno mission will reveal more about Jupiter lightning over the next few years.

References

Aplin, K. L., Fischer, G., Nordheim, T. A., et al. 2020, SSRv, 216, 26

Becker, H., Alexander, J. W., Atreya, S. K., et al. 2020, Natur, 584, 55

Cooper, J. F., Johnson, R. E., Mauk, B. H., Garrett, H. B,, & Gehrels, N. 2001, Icar, 149, 133

Desch, S. J., Borucki, W. J., Russell, C. T., & Bar-Nun, A. 2002, RPPh, 65, 955

Farrell, W. M. 2000, Radioastronomy at Long Wavelengths, ed. R. G. Stone, et al. (Washington, DC: American Geophysical Union)

Giles, R. S., Greathouse, T. K., Bonfond, B., et al. 2020, JGRE, 121, e06659

Goertz, C. K. 1976, JGR, 81, 3368

Greenberg, R. 2008, Unmasking Europa: The Search for Life on Jupiter's Ocean Moon (New York: Copernicus/Springer)

Gurnett, D. A., Kurth, W. S., & Scarf, F. L. 1979, Sci, 206, 987

Hill, T. W., Dessler, A. J., & Goertz, C. J. 1983, in Physics of the Jovian Magnetosphere, ed. A. J. Dressler (Cambridge: Cambridge Univ. Press), 353

Imai, M., Wong, M. H., Kolmašová, I, et al. 2020, GeoRL, 47, e88397

Khurana, K. K., Kivelson, M. G., Vasyliunas, V. M., et al. 2004, Jupiter: The Planet, Satellites and Magnetosphere, ed. F. Bagenal, T. E. Dowling, & W. B. McKinnon (Cambridge: Cambridge Univ. Press)

Pierleoni, C., Morales, M. A., Rillo, G., Holzmann, M., & Ceperley, D. M. 2016, PNAS, 113, 4953

Rinnert, K., Lanzerotti, L. J., Uman, M. A., et al. 1998, JGRE, 103, 22979

Shawhan, S. D. 1979, Solar System Plasma Physics, ed. C. F. Kennel, et al. (Amsterdam: North-Holland), 221

Uman, M. A. 1987, The Lightning Discharge (Orlando, FL: Academic)

Yair, Y., Aviv, R., Ravid, G., et al. 2009, JGRE, 114, D09210

AAS | IOP Astronomy

Electrostatic Phenomena on Planetary Surfaces (Second Edition)

Carlos I Calle and Karen L Aplin

Chapter 12

The Electrostatic Environments of the Outer Planets

12.1 The Electrostatic Environment of Saturn

Saturn is in many aspects very similar to Jupiter. It has an atmosphere composed of about 75% hydrogen and 25% helium with traces of other gases, with clouds of ammonia crystals, ammonium hydrosulfate crystals, frozen and liquid water. Deep in the atmosphere, the hydrogen gas turns into an electrically conductive liquid. As the planet rotates, electric currents are formed in this metallic liquid that in turn generate a magnetic field. The interaction of the solar wind particles with Saturn's magnetic field forms its magnetosphere. NASA'S Cassini mission, which was launched in 1997 and orbited Saturn from 2004 to 2017, has re-defined our understanding of the Saturn system. Its Magnetosphere Imaging Instrument (MIMI) produced an image of Saturn's magnetosphere soon after arrival in 2004 by detecting the hydrogen atoms emitted from regions well outside Saturn's rings (Figure 12.1).

The Voyager 1 mission detected chorus and hiss waves as well as RF signals with its Planetary Radio Astronomy (PRA) instrument, but its imager detected no optical flashes (Desch et al. 2002). The PRA on both Voyager 1 and 2 also detected unusual radio discharges of short duration and narrow band with frequencies ranging from 20 kHz to 40 MHz, the upper frequency limit of the PRA (Farrell 2000). These signals were named Saturn electrostatic discharges (SEDs). These signals originate in the equatorial region of Saturn's atmosphere. In 2004, the Radio and Plasma Wave Science (RPWS) instrument on Cassini detected about 5400 SEDs, which were likely correlated with a possible large equatorial lightning storm lasting several months (Fischer et al. 2006). The difficulty with establishing this correlation was due to the lack of whistler detection. Neither the two Voyagers nor Cassini had been able to detect these dispersive signals, even though they are the source of the observed chorus and hiss waves. Whistler waves propagate along magnetic field lines.

doi:10.1088/2514-3433/ac878ech12

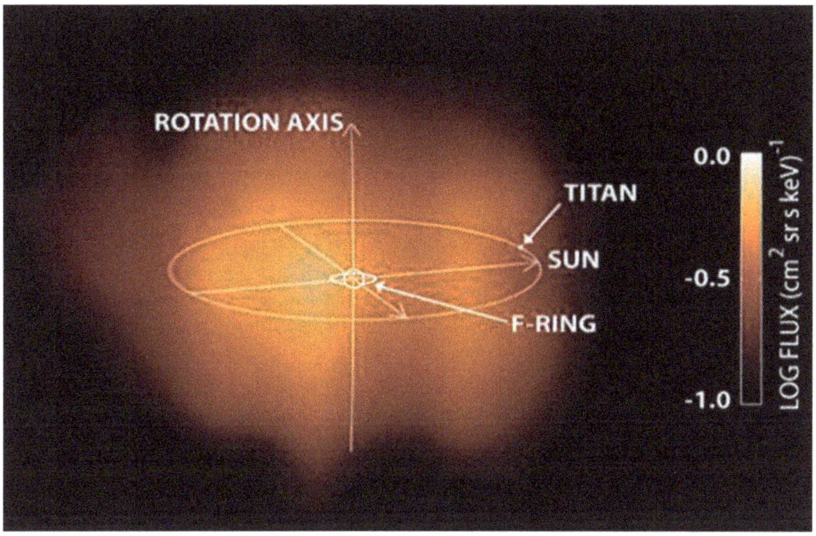

Figure 12.1. Image of Saturn's magnetosphere taken by Cassini's MIMI on 2004 June 21 from a distance of about 6 million km. (Courtesy of NASA.)

Voyager 1 and 2 approached within 3.1 and 2.7 Saturn radii, respectively, thus crossing the magnetic field lines. If the storm was localized in the equatorial region, as suspected, the Voyagers would not have crossed the field lines along which the whistlers traveled (Desch et al. 2002), and Cassini was far outside the magnetosphere and could not detect these signals (Fischer et al. 2006). Corroborating observations of lightning were needed (Aplin & Fischer 2017).

The first visual observation of lightning on Saturn was obtained in August 2009 with Cassini's Imaging Science Subsystem from 35.5 Saturn radii (Dyudina et al. 2010). The images were taken on Saturn's night side, six days after the planet's equinox, when the Sun shone directly over the equator, with the rings edge on to the sunlight which minimized light from the rings shining on Saturn. The optical energy of a single flash was calculated to be 1.7×10^9 J, comparable to that of lightning on Earth and Jupiter (Aplin & Fischer 2017).

Cassini detected daytime lightning on Saturn while observing a giant storm on 2011 March 6 (Figure 12.2). The storm began in 2010 December and lasted until 2011 June. The size of the lightning flashes was measured to be about 200 km in diameter, which was similar to those detected in 2009. From this size, the Cassini mission investigators estimated that lightning occurs 125–250 km below the cloud tops, probably in the water clouds, implying a similar generation mechanism to Earth and Jupiter (Dyudina et al. 2010). The optical energy of single flashes for the 2010–2011 storm was calculated to be 8×10^9 J, somewhat more energetic than the 2009 storms. Radio energies were estimated to be of the same order of magnitude as optical energies.

The total power of the 2010–2011 storm was calculated at about 10^{17} J, which is in the same order of magnitude as the total power that Saturn radiates to space.

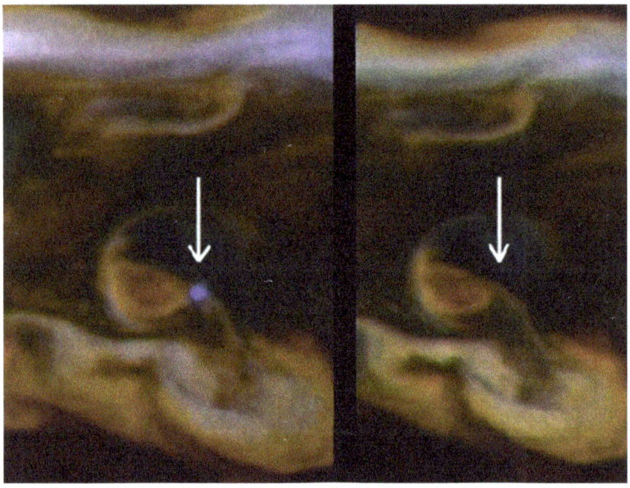

Figure 12.2. False color mosaics from NASA's Cassini spacecraft show lightning strikes during the giant 2010–2011 storm that engulfed the planet's northern hemisphere. The blue dot on the left mosaic is the lightning strike. The mosaic on the right was taken 30 min later and the strike has disappeared. (Courtesy of NASA.)

Ulyana Dyudina and collaborators conclude that these giant lightning storms play an important role in Saturn's cooling and thermal evolution (Dyudina et al. 2013). Radio emissions from Saturn lightning were the first extraterrestrial lightning to be detected from Earth, with a large radio telescope in Ukraine. The storm, in 2006, was simultaneously detected by the Cassini RPWS instrument (Konovalenko et al. 2013).

12.2 The Electrostatic Environments of Uranus and Neptune

Voyager 2 has been the only spacecraft to fly close to Uranus, passing within 81,500 km of the planet in January of 1986. Voyager 2 discovered that Uranus has a strong magnetic field that is inclined at a 59° angle to the planet's axis of rotation and has its center displaced by about 30% of the planet's radius. The inclination of Uranus's magnetic field is larger than those of Saturn, Jupiter, Earth, and Mercury, which are all within about 12° of their axes. Due to this large tilt, the magnetic field lines are wound in a corkscrew shape. The source of Uranus's magnetic field is unknown. No electrically conductive liquid deep in the atmosphere that could produce a magnetic field has been discovered.

Uranus's magnetosphere has a tail that extends to 10 million km. The plasma density of the magnetosphere is very low, about 0.1–1 charged particle (protons or electrons) per cubic centimeter (de Pater & Lissauer 2010). Voyager discovered two radiation belts around Uranus containing mostly protons.

Voyager 2's PWS instrument detected electrostatic discharges from Uranus that resemble those detected from Saturn. These Uranus Electrostatic Discharge (UED) events are not as intense as SED events and are also less frequent. Voyager 2 is also the only spacecraft to visit Neptune, flying within 5000 km of the planet in the summer of 1989. The spacecraft measured Neptune's magnetic field and mapped its magnetosphere. The planet's magnetic field is tilted 47° from Neptune's axis of rotation and is displaced 55% from the planet's geometrical center. Voyager detected auroras over wide regions of the atmosphere of Neptune which had an intensity of about 50% of those on Earth. It also detected whistler waves and a few radio emissions that were much weaker and less frequent than UEDs.

Neptune has a high-altitude cloud layer of methane, a hydrogen sulfide–ammonia (H_2S–NH_3) cloud layer, and deeper layers of water and ammonium hydrosulfide clouds. Lightning could take place in some of these dense cloud layers (Gibbard et al. 1999). Despite the existence of whistlers, there was no optical confirmation of lightning activity. Seran Gibbard and collaborators used a particle-growth and charge separation model that had been applied with great success to both the Earth and Jupiter to investigate the possibility of lightning in the cloud layers of Neptune (Gibbard et al. 1999). With this model, they found that if charge transfer during non-sticking particle collisions can take place faster than 1% of the rate of charge transfer between water ice particles, lightning may happen in the H_2S–NH_3 cloud layer. The model predicted that lightning is inhibited in the deep water and ammonium hydrosulfide cloud layers. Overall, the radio observations suggest that there is lightning at both Uranus and Neptune, but Gibbard's work does not support this theoretically. However, the Gibbard model was limited by many assumptions and lack of data on physical properties of the cloud particles (Aplin et al. 2020). It is also possible that lightning deep within the clouds might not be optically detectable from an orbiting spacecraft.

There may also be electrostatic effects on the clouds at Uranus and Neptune. Wes Lockwood and colleagues at Lowell Observatory spent many years observing these planets with telescopes and concluded that fluctuations in their brightness, resulting from cloud cover, varied with the Sun's 11 year cycle. Two suggestions were proposed for this solar variation. The first idea, unrelated to electrostatics, was that solar ultraviolet radiation was affecting the chemistry of the clouds and changing their color slightly. The second idea, proposed by Julianne Moses and collaborators, was that electrostatic charge released by cosmic rays enhances the formation of cloud droplets. This mechanism is the same as the droplets created in a cloud chamber—it cannot occur naturally in Earth's atmosphere, but may in other planetary atmospheres. As cosmic rays are deflected away from the solar system by the Sun's magnetic field, they also show an 11 year variation. Further analysis of the data obtained by Lockwood, combined with cosmic rays measured by Voyager 2, demonstrated that both mechanisms are involved, so there is likely to be an electrostatic signature in the clouds on both Uranus and Neptune (Aplin et al. 2020).

Uranus and Neptune are the only two planets in the solar system that have not been studied from an orbiting spacecraft, rather than the brief flyby that Voyager 2 offered in the 1980s. NASA is looking into a possible mission to these two planets in the late 2020s or early 2030s. In 2022 it recommended that a mission to Uranus be prioritized. When these missions take place, new data on the cloud structure and other atmospheric properties of these two planets will allow scientists to determine the nature of electrical phenomena in these atmospheres (Mousis et al. 2018).

12.3 The Electrostatic Environment of Saturn's Moon Titan

Titan is Saturn's largest satellite. We include it here with the giant planets because it is part of the Saturnian system and because it is the only satellite in the solar system with a substantial atmosphere. Titan's atmosphere is composed of 96.8% nitrogen, 3% methane, and 0.2% hydrogen with traces of hydrocyanic acid, ethane, propane, acetylene, ethylene, and other gases. The atmosphere has a surface pressure of 1.6 bars. Voyager 1's 1980 flyby did not detect any radio emissions indicative of lightning on Titan (Desch et al. 2002), but knowledge of the composition and meteorology of Titan's atmosphere acquired from Voyager 1 encouraged scientists to suggest that lightning may exist.

In 2001, Tetsuya Tokano and collaborators proposed a model for the generation of lightning on Titan due to cloud electrification during collisions caused by rapid vertical motion of solid methane particles in the atmosphere (Tokano et al. 2001). According to this model, free electrons produced in cosmic ray ionization or impacts from energetic magnetosphere particles attach to methane particles in the cloud. When these negatively charged particles approach Titan's water ice surface, they may form an image charge. This charge separation may generate a maximum temporary electric field of 2.5×10^6 V m^{-1}, large enough to overcome electrical breakdown and produce lightning in the lower atmosphere.

The Cassini mission to the Saturnian system conducted 127 flyby passes of Titan in total, providing new insights into the structure of Titan's upper atmosphere. Despite this thorough investigation, Cassini's RPWS instrument did not detect radio emissions associated with lightning at Titan (Fischer et al. 2020). Titan's deep and hazy atmosphere makes much of it difficult to observe from orbit. To provide the necessary *in situ* observations, the European Space Agency developed a probe called Huygens (after the Dutch astronomer who discovered Titan, and inventor of the grandfather clock, Christiaan Huygens), which was released from NASA's Cassini spacecraft and descended to the surface of Titan in January of 2005. Amongst other instrumentation designed to investigate the atmosphere and surface of Titan, it carried the Huygens Atmospheric Structure Instrument (HASI) to measure atmospheric properties and conductivity, and the Permittivity, Wave, and Altimetry (PWA) instrument to detect radio signals from lightning (Figure 12.3). HASI-PWA detected an unusual ELF signal at 36 Hz (Grard et al. 2006). After several years of analysis of the data collected by these two instruments, Christian Béghin and collaborators argued that this ELF signal displays the characteristics of a second harmonic of a Schumann resonance (Beghin et al. 2012). On Earth, Schumann resonances are generated by lightning in the region (known as a *cavity*) between the Earth's surface and the ionosphere. However, the acoustic sensor on Huygens did not detect thunder that could be associated with lightning, nor was there any other evidence for lightning (Grard et al. 2006). Béghin and collaborators suggested that the interaction of Titan's ionosphere with Saturn's magnetosphere generates electrical currents in the atmosphere, producing a dynamo effect when the magnetosphere plasma rotates with Saturn. The lower boundary of Titan's cavity is proposed to be a conductive water–ammonia ocean residing 55–80 km below an ice crust. These ionospheric current sources and not lightning produce the

Figure 12.3. Artist's rendering of the descent and landing of the Huygens Probe. (Courtesy of NASA/ESA.)

observed Schumann resonance. However, subsequent analysis by Ralph Lorenz and Alice Le Gall suggests that the 36 Hz signal is an artifact, caused by mechanical vibrations in the Huygens system itself (Lorenz & Le Gall 2020).

Cassini and Huygens revealed that Titan is a fascinating world with a rich environment. NASA plans to return there with its Dragonfly mission, planned for launch in 2027 and arrival in 2034. Dragonfly is a robotic rotorcraft carrying a range of instruments to investigate Titan's lower atmosphere and surface.

References

Aplin, K. L., Fischer, G., Nordheim, T. A., et al. 2020, SSRv, 216, 26
Aplin, K. L., & Fischer, G. 2017, Wthr, 72, 46
Beghin, C., Randriamboarison, O., Hamelin, M., et al. 2012, Icar, 218, 1028
Desch, S. J., Borucki, W. J., Russell, C. T., & Bar-Nun, A. 2002, RPPh, 65, 955
Dyudina, U. A., Ingersoll, A. P., Ewald, S. P., et al. 2010, GeoRL, 37, L09205
Dyudina, U. A., Ingersoll, A. P., Ewald, S. P., et al. 2013, Icar, 226, 1020
Farrell, W. M. 2000, Radioastronomy at Long Wavelengths, ed. R. G. Stone, et al. (Washington, DC: American Astrophysical Union)
Fischer, G., Desch, M. D., Zarka, P., et al. 2006, Icar, 183, 135
Fischer, G., Farrell, W. M., Gurnett, D. A., & Kurth, W. S. 2020, JGR, 125, 9
Gibbard, S. G., Levy, E. H., Lunine, J. I., & de Pater, I. 1999, Icar, 139, 227
Grard, R., Hamelin, M., López-Moreno, J. J., et al. 2006, P&SS, 54, 1124
Konovalenko, A. A., Kalinichenko, N. N., Rucker, H. O., et al. 2013, Icar, 224, 14
Lorenz, R. D., & Le Gall, A. 2020, Icar, 351, 113942
Mousis, O., Atkinson, D. H., Cavalié, T., et al. 2018, P&SS, 155, 12
de Pater, I., & Lissauer, J. J. 2010, Planetary Sciences (2nd ed.; New York: Cambridge Univ. Press)
Tokano, T., Molina-Cuberos, G. J., Lammer, H., & Stumptner, W. 2001, P&SS, 49, 539

Ingram Content Group UK Ltd.
Milton Keynes UK
UKHW051607060323
418109UK00006B/40